Lecture Notes in Computer Science 10078

Commenced Publication in 1973
Founding and Former Series Editors:
Gerhard Goos, Juris Hartmanis, and Jan van Leeuwen

More information about this series at http://www.springer.com/series/7409

Anna Satsiou · Georgios Panos
Ioannis Praggidis · Stefanos Vrochidis
Symeon Papadopoulos · Christodoulos Keratidis
Panagiota Syropoulou · Hai-Ying Liu (Eds.)

Collective Online Platforms for Financial and Environmental Awareness

First International Workshop on the Internet
for Financial Collective Awareness and Intelligence, IFIN 2016
and First International Workshop on Internet
and Social Media for Environmental Monitoring, ISEM 2016
Florence, Italy, September 12, 2016
Revised Selected Papers

 Springer

Editors

Anna Satsiou
Information Technologies Institute
Centre for Research and Technology
Thessaloniki
Greece

Georgios Panos
Adam Smith Business School
University of Glasgow
Glasgow
UK

Ioannis Praggidis
Department of Economics
Democritus University of Thrace
Komotini
Greece

Stefanos Vrochidis
Information Technologies Institute
Centre for Research and Technology
Thessaloniki
Greece

Symeon Papadopoulos
Information Technologies Institute
Centre for Research and Technology
Thessaloniki
Greece

Christodoulos Keratidis
DRAXIS Environmental S.A.
Thessaloniki
Greece

Panagiota Syropoulou
DRAXIS Environmental S.A.
Thessaloniki
Greece

Hai-Ying Liu
Norwegian Institute for Air Research
Tromsø
Norway

ISSN 0302-9743   ISSN 1611-3349 (electronic)
Lecture Notes in Computer Science
ISBN 978-3-319-50236-6   ISBN 978-3-319-50237-3 (eBook)
DOI 10.1007/978-3-319-50237-3

Library of Congress Control Number: 2016958994

LNCS Sublibrary: SL3 – Information Systems and Applications, incl. Internet/Web, and HCI

Printed on acid-free paper

This Springer imprint is published by Springer Nature
The registered company is Springer International Publishing AG
The registered company address is: Gewerbestrasse 11, 6330 Cham, Switzerland

Preface

This volume contains the papers presented at the two CAPS workshops co-located with the Third International Conference on Internet Science (INSCI 2016), namely, the First International Workshop on the Internet for Financial Collective Awareness and Intelligence (IFIN 2016) organized by the CAPS PROFIT project, and the First International Workshop on Internet and Social media for Environmental Monitoring (ISEM 2016) organized by the CAPS HackAIR project.

CAPS stands for Collective Awareness Platforms for Sustainability and Social Innovation, and is an EU initiative aimed at designing and piloting online platforms that create awareness of sustainability problems and offering collaborative solutions based on networks (of people, of ideas, of sensors), enabling new forms of social innovation. In line with this initiative, the PROFIT project is about an online platform that aims at promoting the financial awareness and capability of citizens through financial education, financial forecasting, crowdsourcing, recommendations, and engagement tools, in order to enable them to make more informed personal financial decisions, which from an open democracy perspective could also be conducive to more active forms of participation and citizenship. On a different goal, the HackAIR project aims at the environmental awareness of citizens through the development of an online platform that will enable communities of citizens to easily set up air quality monitoring networks and engage their members in measuring and publishing outdoor air pollution levels, leveraging the power of online social networks, mobile and open hardware technologies, and engagement strategies.

The two projects, although targeting different goals, share a lot of common challenges by aiming to develop platforms co-created by European citizens for European citizens toward promoting awareness on different but critical sustainability issues, through the use of crowdsourcing and engaging tools. This volume presents the research works on such and other related technologies and techniques that are being used for the provision of these platforms, targeting the different application domains of financial and environmental awareness, as these were addressed in the IFIN and ISEM workshops, respectively.

The two workshops took place on September 12, 2016, at the University of Florence after the main open session of the INSCI conference, where each project and respective workshop were introduced by their representatives.

September 2016

Anna Satsiou
Georgios Panos
Ioannis Praggidis
Stefanos Vrochidis
Symeon Papadopoulos
Christodoulos Keratidis
Panagiota Syropoulou
Hai-Ying Liu

Contents

IFIN 2016

Preface

The main purpose of IFIN 2016 was to open a multidisciplinary dialogue on how we could use the Internet to promote financial awareness and capability among citizens. More specifically, the new socioeconomic landscape that has prevailed in the post-crisis world brought changes in the financial, labor, and pension markets along with changes in public finance decisions and the political sphere. However, the volume of information coming from the Web, the existence of often ad hoc sources, the documented existence of cognitive limitations by individuals when it comes to the processing of large volumes of information, and the documented widespread financial illiteracy even within developed economies, including those of the European Union, all identify the need for appropriate methods and tools to extract and process such information, create new knowledge and present it to the users.

In this context, IFIN 2016 consisted of two parts: (a) presentation of recent works in the area of financial collective awareness and intelligence including a keynote speech by Prof. Steve Schifferes, and (b) a roundtable interactive discussion with all workshop participants.

Professor Steve Schifferes is currently Director of the Financial Journalism MA at City University London, and has a wide-ranging background in business and finance journalism, both for television and online. He has been economics correspondent for BBC News Online, coordinating coverage of the credit crunch, the Asian financial crisis, the Enron scandal, and the launch of the euro. His keynote speech focused on the challenge of financial literacy and the Internet's potential to improve people's ability to function well in the financial system. He began by defining financial literacy, which requires understanding of basic financial concepts, e.g., interest rates, inflation, GPD, etc., and is a pre-condition for financial capability. According to Prof. Schifferes, a substantial proportion of the public (30–40%) has limited financial literacy, and typically women, young people, and poorer people score lower on financial literacy. Moreover, he addressed the dramatic effect that the Internet has had on the financial sector. As it was mentioned, it has accelerated the pace of financial transactions and offered instant access to up-to-date financial information, including personal bank accounts, but at the same time, it has also made it easier for false rumors that affect financial markets to spread rapidly, and it has perhaps made it more difficult for individuals to distinguish reliable from unreliable financial advice. Financial institutions are already making extensive use of data mining and tracking both to target individuals and to predict market movements, and automated trading and dark pools are making market pricing less transparent. Meanwhile, the authorities have been slow to respond to many of these developments, which are being led by private companies, and, today, there is also less public debate about policy and less trust from citizens in decision-makers.

According to Prof. Schifferes, the Internet can play a major positive role in addressing needs in this area, thanks to:

- Tools that improve financial literacy, including games, education tools, and assessment tools
- Interactive automated counseling systems to help individuals examine their options regarding debt investment and life course choices of a financial nature
- Improvements in the way the media handles, displays, and explains financial data and a stronger emphasis on financial literacy and less use of jargon
- Experiments in peer-to-peer counseling and recommendation system
- Better understanding of the working of commercial advice and recommendation systems such as declaring commissions

Concluding, it is important that we both understand, oversee, and contribute (especially as researchers) to these developments, since the current economic climate requires that individuals become responsible for their own financial decisions, without the support of the state, and, therefore, both access to financial information and financial literacy are deemed of utmost importance.

The following multidisciplinary research works that were presented at the IFIN worskhop address many of the various important issues raised by Prof. Schiffere's keynote speech and are presented in the following sections of this book along with the conclusions from the roundtable discussion.

- "What Do People Expect from a Financial Awareness Platform? Insights from an Online Survey." This paper presents the results of the questionnaire that was distributed to the wider public before the design and development of the PROFIT platform. The questionnaire aimed at evaluating target user's financial behavior, knowledge, awareness, and capability as well as their attitude toward technology and Internet use, their suggestions for a financial awareness platform's material and functionalities, and their potential motives for participating in and contributing to a financial awareness platform, such as the PROFIT platform.
- "A Reputation-Based Incentive Mechanism for a Crowdsourcing Platform for Financial Awareness." The contribution focuses on a proposed reputation-based mechanism for encouraging and sustaining user participation in the PROFIT platform, which was designed taking into account both the best practices used in other crowdsourcing platforms, as well as the results from the online questionnaire distributed previously to the wider public.

- "Predicting Euro Stock Markets." In this paper, the authors attempt to forecast the sign reversals of the Eurostoxx 50. Industrial portfolios are used as predictive variables as well as an oil sentiment index constructed using textual analysis. Results indicate that both the industry portfolios and the oil sentiment index have some predictive power in a one-month period ahead.
- "On the Quality of Annotations with Controlled Vocabularies." This work investigates how one can prepare a controlled vocabulary in the form of a thesaurus, based on the analysis of a high-quality corpus, as well as qualitative measures describing the usefulness of articles when searching for new candidate concepts for inclusion. The introduced approach is tested with two financial thesauri and corpora of financial news.

The editors would like to thank all the authors for submitting their work, the Program Committee members for the high-quality reviews they provided, as well as the INSCI 2016 organizers for their constant support.

We hope that these proceedings will be a source and reference for further fruitful discussions and research works in the area of financial awareness.

September 2016

Anna Satsiou
Georgios Panos
Ioannis Praggidis

Organization

General Chairs

Anna Satsiou — Information Technologies Institute, Centre for Research and Technology Hellas, Greece

Georgios Panos — Adam Smith Business School, University of Glasgow, UK

Ioannis Praggidis — Department of Economics, Democritus University of Thrace, Greece

Program Committee

Nana Oiza Akubulem — University of Glasgow
Christian Blaschke — SWC
Christoforos Bouzanis — University of Glasgow
Dimitrios Chronopoulos — University of St. Andrews
Anastastios Economides — University of Macedonia

Dimitrios Gounopoulos	University of Newcastle
Maria Grydaki	University of Stirling
Frank Hopfgartner	University of Glasgow
Eirini Karapistoli	Democritus University of Thrace
Katerina Katmada	CERTH-ITI
Ioannis Kompatsiaris	CERTH-ITI
Alexandros Kontonikas	University of Glasgow
Sergei Kuznetsov	NRHSEU
Frank Hong Liu	University of Glasgow
Emmanuel Mamatzakis	University of Sussex
Apostolos Mavridis	Aristotle University of Thessaloniki
Stuart Middleton	University of Southampton
Alberto Montagnoli	University of Sheffield
Mirko Moro	University of Stirling
Evangelos Vagenas-Nanos	University of Glasgow
Theophilos Papadimitriou	Democritus University of Thrace
Symeon Papadopoulos	CERTH-ITI
Vasilios Plakandaras	Democritus University of Thrace
Artem Revenko	SWC
Panagiotis Sarigiannidis	University of Western Macedonia
Steve Schifferes	City University London
Georgios Sermpinis	University of Glasgow
Vasilios Sogiakas	University of Glasgow
Panagiotis Tsintzos	Democritus University of Thrace
Michail Vafopoulos	IIT-Demokritos
Stefanos Vrochidis	CERTH-ITI
Robert Wright	University of Strathclyde

Supported by

Horizon 2020 PROFIT project.

PROFIT
Promoting Financial Awareness and Stability

IFIN 2016 Roundtable Discussion

The IFIN 2016 roundtable discussion was led by Georgios Panos, Professor of Finance in the Adam Smith Business School (University of Glasgow). Prof. Panos, a financial and labor economist, initiated the discussion by providing the attendees with food for thought. More specifically, he presented the following text to the participants and asked them to write in a paper the first word that came to their minds when reading this text:

In X country, the financial wing of Y company, with 400 million customers, is constructing a credit scoring system. Its scoring algorithm seemingly relies on information beyond financial transactions, augmenting it with data from social media interactions to construct what is described as a "social credit score". The context of these developments is a government initiative to construct a national database of social credit scores for all citizens by 2020. Credit scores will be based on a "complete network covering credit records" and will serve to establish "the idea of a sincerity culture, and carrying forward sincerity and traditional virtues." This is a large-scale and ambitious social experiment.

In summary, the idea that was provided to participants to ponder upon was the creation of what was called a social credit scoring system in X country by the financial wing of Y company, one of the world's biggest online shopping platform with 400 million customers. In the general case, a credit score system is used to predict the creditworthiness of an individual based on credit report information sourced from credit bureaus. However, in the particular text provided to the participants a new social credit score was mentioned that is also based on data from social media interactions, as part of a government initiative, presented as an ambitious social experiment. The participants were asked to mention the first word that came to their minds regarding this so-called large-scale social experiment. It was very interesting to see that the participants' first impressions on this matter varied widely from totally negative to totally positive irrespective of their background. The negative characterizations of this idea were: risky, controlling, authoritative, Foucault-ian, nightmarish, a panopticon, and scary, while the positive characterizations were interesting, bold, ambitious, innovative, and challenging.

As the participants were later informed, this credit score is based on information extracted from the Y company's consumer information database. In particular its scoring algorithm seems to rely on the financial and consumption activities of its users, augmented with data from their payment, purchases, and social media interactions to construct a social credit score, which users are even encouraged to showcase on their social media. Apart from Y company, several other companies of X country also work on social credit scoring systems. Given the fact that information of that kind is not readily available or easy to measure in X country, such a social credit score system could facilitate its measurement. On the other hand, attendees were concerned regarding the privacy of the citizens' credit scores and whether Y companys users are voluntary and knowingly rated by the social credit score. The negative future repercussions that a bad social score could have on an individual's life are something that everyone agreed should be taken into serious consideration. These should also lead us to rethink the serious social and ethical consequences of online platforms in general and financial Web platforms in particular, as well as the use of social media in the personal finance domain. More importantly, we should ask how far we are willing to go and boundaries should be delineated. Thus, the controversy of this bold example, in conjunction with the prominent heterogeneity of the audience as regards their work/education background and their views on the matter, led to a very interesting discussion.

A concluding remark from the discussions is that information should be put into context and perspective; more importantly, we should clarify what type of information is actually useful and necessary inside the platform, and what type outside. Cultural differences, the cultural context, as well as differences based on the user group (e.g., user age) should be taken into consideration, in any case. The aforementioned paradigm, e.g., takes place in a particular country. Nevertheless, this country does not have a national and widely used credit system like those implemented in Western countries. Therefore, it is argued that private lenders like Y company have created such systems to better inform their lending decisions. Nevertheless, there are norms at an international level for how data should be collected and used for calculating, e.g., credit scores. Legal and ethical issues should be taken into consideration. Therefore, important information that should be provided to users is how the personal information they may provide in a financial awareness platform is going to be used, who is going to use it, and whether it will be used in order to create a new product or for marketing purposes.

The role and importance of financial education were also mentioned. More specifically, we should not forget the importance of: (a) personal contact in financial education, and the fact that people still rely a lot on advice from friends, family, etc. for financial issues, and (b) a financial awareness platform such as the PROFIT platform should on one hand provide users with information from various sources in order to broadly inform them and provide them with a spherical vision, and on the other hand provide them with the tools and knowledge to assess the information, interpret the data in the right way, and use the provided material in a rational way. Concerning, e.g., financial forecasting, it was argued that we should provide users with information on what financial forecasting actually is, making them that way more aware and more cautious about the information provided. Thus, going a bit further, it could be argued that a social credit score system like the aforementioned one and like any other such innovative measure should be presented to the public along with all the information needed for people to understand how this is calculated, how it is being used, the negative effects it may have on their future decisions, as well as the importance of keeping a good credit history.

Another issue that arose in the roundtable discussion is how we can maximize and measure the social impact of a financial awareness platform. A prerequisite would be, of course, to reach and engage large numbers of users. A good product, as well as collaborations with institutions would be useful; especially, partnerships with financial consumer protection agencies could act very positively. However, the platform's social impact should be measured not only in terms of the number of people it reaches but most importantly in terms of attitude change of these people toward more responsible personal financial decisions. In order to be able to evaluate such behavior change, much care should be taken to identify the appropriate metrics and ways to do it, and this is another challenge for the PROFIT project.

What Do People Expect from a Financial Awareness Platform? Insights from an Online Survey

Georgios A. Panos[1], Konstantinos Gkrimmotsis[1],
Christoforos Bouzanis[1], Aikaterini Katmada[2(✉)], Anna Satsiou[2],
Gian-Luca Gasparini[3], Aurora Prospero[5], Ioannis Praggidis[4],
and Eirini Karapistoli[4]

[1] Adam Smith Business School, University of Glasgow, Glasgow, UK
{Georgios.Panos,Konstantinos.Gkrimmotsis,
Christoforos.Bouzanis}@glasgow.ac.uk
[2] Information Technologies Institute, CERTH, Thessaloniki, Greece
{akatmada,satsiou}@iti.gr
[3] SEFEA Consulting, Padova, Italy
g.gasparini@sefea.org
[4] Democritus University of Thrace, Komotini, Greece
{gpragkid,ikarapis}@econ.duth.gr
[5] FEBEA, Brussels, Belgium
a.prospero@sefea.org

Abstract. The aim of this study is to present the analysis of an online survey that was conducted in order to investigate individual attitudes and requirements from an online financial awareness platform. The survey aimed to elicit users' self-assessed financial knowledge, financial capability and awareness, along with facets of their financial behaviour. Moreover, it entailed questions capturing attitudes towards technology and internet usage. Specifically, it targeted requirements for specific resources and features of a financial awareness platform, along with explicit motivations and incentives for participating and contributing to the platform of the PROFIT project. The custom-made online survey was completed by 494 respondents from different demographic groups and user groups, i.e., in terms of familiarity and requirements. The results indicate that there is a strong existing need in the market for online financial information and awareness development with online tools.

Keywords: Financial awareness · Online platform · Online survey · PROFIT

1 Introduction

The recent financial crisis has generated interest among citizens in better following and understanding economic matters and financial trends. Moreover, from an institutional and social constructionist perspective, there is also a great renewed interest on how to promote more responsible individual saving and borrowing behaviour. The ability of citizens to make informed financial decisions is critical to developing sound personal

© Springer International Publishing AG 2016
A. Satsiou et al. (Eds.): IFIN and ISEM 2016, LNCS 10078, pp. 9–56, 2016.
DOI: 10.1007/978-3-319-50237-3_1

finance, which can contribute to increased saving rates, more efficient allocation of financial resources, and greater financial stability [18].

Technological developments have enabled and enhanced the availability of large volumes of information on themes relevant to financial decision making. Their potential benefits, though, are hindered by the cognitive limitations by individuals when it comes to the processing of large volumes of information, as well as the documented widespread financial illiteracy even within developed economies, including those of the European Union.

Acknowledging such needs, the PROFIT project [1], engaging information and communications technology, along with financial experts and social partners, will develop a financial awareness platform built upon Open Source components. The PROFIT Platform is conceived aiming to cater to the recognized need of action enhancement for greater financial awareness and capability. Financial knowledge has been identified as a major target for improved social performance, client protection, and, ultimately, greater societal well-being. More specifically, the PROFIT platform will be a user-centered financial awareness platform. It is going to obtain finance-related crowdsourced data from the web as well as its users, and create new knowledge aiming to offer financial information, education, and advanced forecasting tools to help users understand financial data and trends. That way, it aims at empowering them in decision-making and financial capability, catering to their specific profile, and filling the existent gap among other analogous multi-functional software solutions in the European environment.

Overall, the objectives of the PROFIT platform are the following: (a) to raise financial awareness and support better decision-making; (b) to create financial collective intelligence; and (c) to empower user participation and interaction. In order to achieve such objectives, the first part of the project would be the "identification phase". It constitutes the beginning of the project and involves the identification of user requirements. The effort was initiated via the design of a custom-made survey, which was implemented online aiming to acquire information regarding participants' financial behaviour and interests as potential users of the PROFIT platform. The survey involved an online questionnaire[1] of some 35 questions, which was available to the wider public and translated in six languages. The analysis of the responses to this questionnaire provided useful information regarding the features and functionalities that could be implemented in the platform, as well as the needs and interests of the participants.

It was interesting to see from the online survey results that most respondents do not know or answer when it comes to their source of information for changes in financial markets. Moreover, it was also found that the web is the third most frequent source of financial information and advice prior to purchase. The discrepancy between the high figures for those who do not know how to seek for information regarding changes in financial trends and services, and the notable rather low figure for the use of specialized websites indicates that a financial awareness platform, such as the PROFIT platform, is

---

[1] The realization of the PROFIT's online questionnaire has received the ethical approval from the Bio-Ethics Committee of the Centre for Research and Technology Hellas with REF No: ETH. COM-19.

likely to cater to a strong existing need in the market for financial knowledge and information. The positive effects of financial knowledge are quite significant and *iii lilii iimprovement* of wealth management, life planning, and enhanced levels of loan repayment. In general, citizens can *mili* *iiiiii* informed personal financial decisions which, from an open democracy perspective, could be conducive to *more active formii* of participation and citizenship. All the above dimensions can be directly and indirectly linked to enhanced financial stability.

The structure of this study is as follows: Sect. 2 reviews the relevant literature that led to the specification of user groups for the PROFIT project. Then, Sect. 3 presents the analysis of the responses to our online survey. Finally, Sect. 4 discusses the main findings and draws some conclusions pointing to the future directions of the generation of a financial awareness platform, such as the PROFIT platform.

2 Literature Review

Based on the growing literature on financial literacy, the initial proposal by the experts in the PROFIT consortium was to identify 17 distinct user profiles. More specifically, these comprised of the following: (1) Entrepreneurs/latent entrepreneurs; (2) Students; (3) Elderly/retirees/pre-retirees; (4) immigrants/members of ethnic minorities; (5) females; (6) children and parents of young children; (7) insurance clients; (8) clients from indebted and over-indebted households; (9) investors and potential investors; (10) the unemployed; (11) active citizens; (12) mortgage owners/home owners/first-time buyers; (13) employees/trainees; (14) professionals in financial services; (15) government executives and political party members; (16) media operators; and (17) taxpayers.

The proposal was based on modern and recent evidence. Indeed, Klapper, Lusardi, and Panos [2] argue that entrepreneurs in the US are found to be some 25% more financially literate compared to non-entrepreneurs. Moreover, among entrepreneurs, the more financially literate perform better in a number of indicators related to entrepreneurial performance. According to Brown et al. [3], variation in exposure to financial training among US students comes from state-wide changes in high school graduation requirements. Using a flexible event study approach, it was found that both mathematics and financial education decrease reliance on non-student debt and improve repayment behavior. Economics training, on the other hand, increases both the likelihood of holding outstanding debt and the prevalence of repayment difficulties. Furthermore, Panos and Wright [4] find that among Scottish students, gender differences in financial literacy start very early on life and economics training can account for the largest unobserved part of gender discrimination in terms of financial literacy.

With respect to the elderly/retirees and pre-retirees, a large body of evidence shows that baby-boomers in the US are among the most challenged groups in terms of financial-literacy skills (e.g. [5]). Moreover, given a certain cognitive decline that occurs with old age and a certain lack of technological literacy, one can imagine why the elderly are a group of particular interest. Recent evidence shows that financial knowledge is a key determinant of wealth inequality at old age. 30–40% percent of retirement wealth inequality in the US is accounted for by financial knowledge [6]. Demirguc-Kunt, Klapper and Panos [7] show that the incidence of saving for old age is particularly low around the world, i.e. 25% of individuals around the world save for old age, with a figure close to 35% for high-income OECD countries. Hence, this is a group of particular interest. Lusardi and Mitchell [8], Klapper and Panos [9] and van Rooij et al. [10] show that financially-literate individuals around the world are more likely to prepare for retirement, and more likely to engage in retirement planning using private pension plans.

Moreover, Lusardi and Mitchell [5] suggest that certain vulnerable population groups, including the immigrants, appear to be lacking in terms of financial literacy. Given the very high volumes of remittances from abroad around the world for immigrant families, the existence of informal financial transfer networks for remittances, and the prevalence of digital networks for remittances, one can easily assert why low financial literacy skills among immigrants can bring further disadvantage to a financially-constrained population group in an era of digitized payments. Worldwide evidence also documents large gender differences in terms of financial literacy [5]. It has been suggested often that the gender differential can in part be explained by differences in confidence. There is a large body of literature suggesting that males tend to be overconfident regarding their financial skills (e.g. [11]). Further evidence [4] indicates that gender differences in financial literacy start very early on in life, i.e., from the age of 10. Moreover, economics/business training can account for the largest part of the unobserved discrimination in financial literacy between male and female students in Scotland. Moreover, again examining Scottish students, McPhail et al. [13] show that financial training at home by parents and at school by teachers both play a pivotal role in shaping student financial awareness.

Recent literature on behavioral insurance [14] suggests that financial literacy and financial advice can mitigate the behavioral temptation to lapse into surrendering life insurance policies early, while the tendency to rely on heuristics increases the lapse probability. Given that many policyholders surrender their life insurance policies early, leading to substantial monetary losses for private households, the financial literacy skills of potential life insurance customers are of particular interest as a distinct user case. Lusardi and Tufano [15] suggest that the least financially savvy incurred high transaction costs, paying higher fees and using high-cost borrowing. In their study, the less knowledgeable also reported that their debt loads were excessive, or that they were unable to judge their debt positions. Since then, there has been follow-up evidence, reviewed in [5] to suggest that low financial literacy is linked with the prevalence of over-indebtedness among households (e.g. [16], inter alia). Studies in United States and elsewhere have found that the more financially literate are also more likely to participate in financial markets and invest in stocks (e.g. [17], inter alia).

With respect to the unemployed being considered a distinct user group, there is evidence in many studies suggesting that the unemployed rank low in terms of financial literacy in several countries (e.g. [18–20]). Moreover, the inference that perhaps financial literacy can be seen a useful part for adult and lifelong learning stems from the job-tasks approach in modern labour economics. In a seminal paper, Yuamaguchi [21] proposes a heterogeneous human capital model using task data from the Dictionary of Occupational Titles for the US. His novel use of task information enables a clear distinction between worker skills and job tasks. Summarizing the information on task complexity across 62 characteristics, the author constructs a complexity measurement for two broadly defined tasks: cognitive tasks and motor tasks. Gathmann and Schönberg [22] propose the concept of task-specific human capital, using German data וווו ו וווא וririzing 19 tasks for 64 occupations into three aggregate groups: analytical tasks, manual tasks and interactive tasks. In similar spirit, Autor and Handel [23] categorize 91 occupations in the US into three aggregate domains, cognitive, והוו personal and physical activities. Evidently, financial literacy is a key cognitive skill that can be seen as highly relevant.

In terms of open democracy and the formation of attitudes towards public affairs, public economics and finance by active citizens, recent evidence suggests that financial literacy is of great relevance. Salient political choices – recently involving voting in referendums – are determined by attitudes towards redistribution, immigration, austerity, as well as the working of economic partnerships and monetary and fiscal unions etc. Such attitudes are likely to depend upon the understanding of the basics of macroeconomic accounts and public finance. Recent evidence indicates that there are relationships between financial literacy in the UK and attitudes to redistribution, the Scottish referendum, the EU referendum, and towards immigration (e.g. [19, 20, 24, 43]). Research also shows that financial knowledge is a key determinant of wealth inequality. 30–40% percent of retirement wealth inequality in the US is accounted for by financial knowledge [7].

With respect to home ownership/first-time ownership and mortgages, Lusardi and Mitchell [5] review how the large number of mortgage defaults during the financial crisis has suggested that debt and debt management is a fertile area for mistakes. For instance, many borrowers do not know what interest rates were charged on their credit card or mortgage balances (e.g. [25–27]). Thus, first-time owners, mortgage owners and home owners can be thought to make particular cases for a user group. With respect to employees/trainees, the rationale is similar to that for the unemployed in (10). Moreover, financial literacy has also been linked to the demand for on-the-job training [28] and being able to cope with financial emergencies [29].

Professionals in financial services are a particular group of interest. Bodnaruk and Simonov [30] analyze private portfolios of mutual fund managers. They find no evidence that financial experts make better investment decisions than peers: they do not outperform, do not diversify their risks better, and do not exhibit lower behavioral biases. However, managers do much better in stocks for which they have an information advantage over other investors, i.e., stocks that are also held by their mutual funds. The rationale for considering government executives and political party members as a distinct user group is similar to that for active citizens (11), as for the quality of public dialogue to be upgraded, particularly that which involves economic and financial matters, there needs to be a standard level of understanding and being able to communicate over basic financial and economic notions.

The rationale for considering media operators as a distinct user group of interest is also similar. There is a clear link between the dissemination of information and public choice, and between media markets and policy making (see, e.g., the essays in Islam [41]). For that link to be transparent and based on accuracy there needs to be a standard level of financial literacy for awareness to prevail among the public. Lastly, for taxpayers, it is obviously the case that financial literacy is positively related to making informed choices, both in terms of own expense accounts and in terms of choosing a financial advisor [42].

Upon deliberation among experts, social partners and users, in the following section we analyse the online survey results for the 12 most relevant to PROFIT aforementioned user profiles, namely (1) Entrepreneurs/latent entrepreneurs/social entrepreneurs/self-employed, (2) Elderly/retirees/pre-retirees, (3) Migrants/Members of an ethnic minority, (4) Parents of young children, (5) Customers: Indebted/Overindebted households, (6) Customers: Investors/Potential investors/Depositors, (7) Unemployed/trainees, (8) Active citizens/taxpayers, (9) Mortgage owners/home owners/first-time buyers, (10) Professionals in financial services/financial experts, (11) Government executives and political-party members/local authorities, (12) Collective investors/borrowers/third sector organisations, as these have been decided together with the PROFIT User Forum Committee that includes representatives of all key stakeholders (financial institutions, entrepreneurs, government bodies, educational institutions, banks' customers and other potential customers and/or final users of the PROFIT platform), as well as high-profile experts in the area.

3   The PROFIT Online Survey

This section presents the findings from the analysis of the 494 responses received to our online survey. The aim is to understand the characteristics and user requirements of potential users of the PROFIT platform. A custom-made questionnaire was implemented capturing at least six sets of characteristics, presented in the following sub-sections. These can be distinguished between: (a) demographic characteristics; (b) financial behavior; (c) financial knowledge; (d) financial sentiment, news and informational retrieval; (e) financial literacy training requirements; and (f) technological literacy and incentive compatibility requirements.

The questionnaire involved 35 questions. It was designed by experts at the PROFIT project in early 2016. The translation and the creation of the online version of the questionnaire both took place in the spring of 2016. Data collection was carried out in the summer of the same year. The detailed content of the English version of the questionnaire can be found in http://projectprofit.eu/material/#tab-id-5. Overall, there were 494 respondents, 248 of which (50.2% of the sample) filled the English version of the questionnaire, with the remaining 246 (49.8%) filling versions in other languages. Specifically, 30 respondents (6.1%) filled the Italian version of the questionnaire, 120 (24.3%) filled the Greek version, 35 (7.1%) filled the Croatian version, 33 (6.7%) filled the Slovenian version and 28 (5.7%) the French version.

In the following sub-sections, the summary statistics of the responses are being explained and illustrated in figures. In the majority of the cases, these are presented: (i) for the pooled sample overall, (ii) by English-response status, and (iii) by self-assessed user type. The user types are identical to the twelve groups that were identified in the previous section.

The initial design targeted 18 users per user type, i.e., 12 user profiles*18 users per profile = 216 respondents in total. The 494 responses received exceed the minimum requirement. However, the minimum requirement was not reached with respect to four

Table 1. User type distribution

	(1) All	(2) English	(3) Non- English	(4) Diff.	
UG₁	18.3%	18.4%	19.7%	0.0990	***
UG₂	14.4%	21.8%	6.9%	0.1811	***
UG₃	1.2%	0.8%	1.6%	-	
UG₄	6.1%	7.7%	4.5%	0.0433	
UG₅	1.6%	1.6%	1.6%	-	
UG₆	6.3%	8.1%	4.5%	-0.0194	
UG₇	5.1%	1.2%	9.0%	0.0121	***
UG₈	32.1%	29.8%	34.3%	0.1317	
UG₉	7.7%	12.5%	2.9%	-0.0083	***
UG₁₀	5.7%	5.2%	6.1%	-0.0809	
UG₁₁	0.6%	0.4%	0.8%	-	
UG₁₂	1.0%	0.0%	2.0%	-	

		(5) All	(6) English	(7) Italian	(8) Greek	(9) Croatian	(10) Slovenian	(11) French
ALL		494	248	30	120	35	33	28
UG_1	Entrepreneurs/latent entrepreneurs/ social entrepreneurs/self-employed	90	27	3	43	10	5	2^t
UG_2	Elderly/retirees/pre-retirees	71	54	2^t	2^t	1^t	3	9
UG_3	Migrants/Members of an ethnic minority	6^t	2^t	3	0^t	0^t	1^t	0^t
UG_4	Children/Parents of young children	30	19	1^t	5	2^t	2^t	1^t
UG_5	Customers: Indebted/Overindebted households	8^t	4	1^t	0^t	0^t	3	0^t
UG_6	Customers: Investors/Potential investors/Depositors	31	20	3	3	0^t	0^t	5
UG_7	Unemployed/trainees	25	3	0^t	14	3	5	0^t
UG_8	Active citizens/taxpayers	158	74	5	44	12	12	11
UG_9	Mortgage owners/home owners/ first-time buyers	38	31	4	1^t	1^t	1^t	0^t
UG_{10}	Professionals in financial services/ financial experts	28	13	4	8	3	0^t	0^t
UG_{11}	Government executives and political-party members/local authorities	3^t	1^t	1^t	0^t	1^t	0^t	0^t
UG_{12}	Collective investors/borrowers/third sector organisations	5^t	0^t	3	0^t	2^t	0^t	0^t

Notes: The symbol t indicates categories in which the minimum target of 3 respondents (18 overall) was not reached. Asterisks denote the following levels of significance from unpaired t-tests for mean differences: * p<0.10, ** p<0.05, *** p<0.01. Differences for groups 3, 5, 11, and 12 are not reported or tested, as the sample sizes are small and unsuitable for parametric statistics user profiles, namely the migrants, the over-indebted, government executives, and collective investors/borrowers/3rd sector. Evidently, these are user types that were difficult to attract to the online survey. Hence, these are also the user groups, for which inferences cannot be easily drawn from this survey.

Table 1 presents the distribution of the specified user types. Panel A presents sample averages overall (column 1) and by English language status (columns 2–4), respectively. Differences between the respective groups, along with the statistical significance, are shown in column 4. Then, Panel B presents the number of responses obtained per user group and language. The inspection of the table indicates that 18.3% of the sample (90 respondents) is in the user group 1 (*Entrepreneurs/latent entrepreneurs/social entrepreneurs/self-employed*). 14.4% (71 respondents) are in the user group 2 (*Elderly/retirees/pre-retirees*). Only 1.2% of the sample (6 respondents) are in the user group 3 (*Migrants/Members of an ethnic minority*). 6.1% (30 respondents) belong to the user group 4 (*Parents of young children*). Only 1.6% (8 respondents) belongs to the user group 5 (*Customers: Indebted/Over indebted households*). Furthermore, 6.3% (31 respondents) belong to the user group 6 (*Customers: Investors/Potential investors/Depositors*) and 5.1% (25 respondents) belong to the user group 7 (*Unemployed/trainees*). Interestingly, the biggest self-defined type is the user group 8 (*Active citizens/taxpayers*), with 32.1% of the sample (158 respondents). In addition, 7.7% (38 respondents) belong to the user group 9 (*Mortgage owners/home owners/first-time buyers*) and an additional 5.7% (28 respondents) belong to the user group 10 (*Professionals in financial services/financial experts*). The targets were not reached with respect to the last two user types. Only 0.6% (3 respondents) belong to the user group 11 (*Government executives and political-party members/local authorities*) and an additional 1% (5 respondents) belong to the user group 12 (*Collective investors/borrowers/third sector organizations*).

Overall, 3 user groups were not identified among English respondents, 5 user groups were not identified among Italian respondents, and 6 groups were not found among Greek respondents. In addition, 8 user groups were not identified among Croatian respondents, 7 groups were not shown among Slovenian respondent and 9 groups were not shown among French respondents.

English respondents are significantly more likely to be found among user groups 1, 2, and 7 (entrepreneurs, the elderly, and unemployed). They are less likely to be found in user group 9 (mortgage owners). It is worth noting that differences for groups 3, 5, 11, and 12 are not reported or tested, as the sample sizes are small and unsuitable for parametric statistics.

3.1   Demographic Characteristics

Figure 1 presents the gender distribution in the sample and across sub-sample and user types. 59.9% of the respondents in the sample are males, with the remaining 40.1% being females. 61.7% of the English respondents in the sample are males, compared to 58.1% of the non-English sample. However, the difference between these two groups is not statistically significant. The inspection of the remaining bars in the figure suggests that there are more males among the entrepreneurs (UG1: 74.4%), the elderly (UG2:

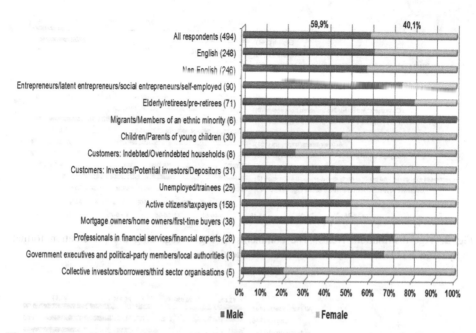

Fig. 1. Gender distribution. *Notes*: Distributions by gender are presented (i) for the pooled sample of all respondents, (ii) by English/non-English response questionnaire, and (iii) by user group type. Gender is obtained from question 2 in the questionnaire.

80.3%), the potential investors (UG6: 64.5%), the professionals in financial services (UG10: 71.4%). Females are more likely to be found among parents of children (UG4: 53.3%), the unemployed (UG7: 56%) and mortgage owners (UG9: 60.5%). Again, differences for groups with small numbers of responses are not reported, e.g. all 6 migrant respondents are male, etc.

The age distribution of the sample and sub-samples is shown in Fig. 2. The first bar in the figure indicates that 8.9% of the sample is aged between 18 and 25, 25.3% of the sample is aged 26–35, and 22.9% of the sample is aged 36–45. Moreover, 16.8% if the sample is aged between 46 and 55, 15.2% is aged between 56 and 65, and 9.5% is in the group aged between 66 and 75. There is a remainder 1.4% of individuals in the sample who are more than 75 years old. Comparison of the remaining bars in the figure indicates that English language respondents are much older, compared to the non-English language respondents. Specifically, the average age among English respondents is 49.7 years; compared to 38.7 years among non-English respondents (these averages are not shown). The average age is notably higher among the elderly (UG2) and the potential investors (UG6), and notably lower among entrepreneurs (UG1), parents of children (UG4), the unemployed (UG7), active citizens (UG8), mortgage owners (UG9) and professionals in financial services (UG10).

Figure 3 presents the educational distribution of the sample. It is evident that the average number of years of education in the sample is high (16.9 years). Moreover,

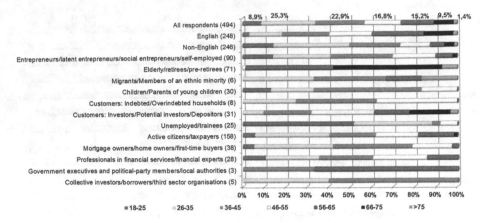

Fig. 2. Age distribution. *Notes*: The comments in Fig. 1 regarding the presentation format apply. Age is obtained from question 3 in the questionnaire.

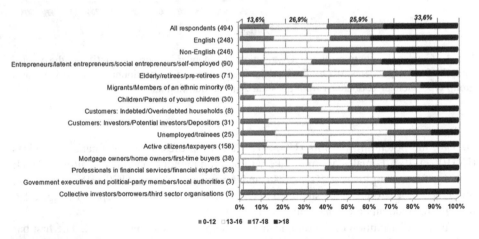

Fig. 3. Education distribution. *Notes*: The comments in Fig. 1 regarding the presentation format apply. The years of education are grouped in the four categories above: (i) 0–12, (ii) 13–16, (iii) 17–18, and (iv) >18. Years of education are from Question 4 in the original questionnaire.

13.6% of the respondents have 0–12 years of education, with the minimum figure being 2 years. 26.9% of the respondents have 13–16 years of education. 25.9% of the individuals in the sample have 17–18 years of education and the remainder 33.6% of the sample have more than 18 years of education. Thus, it is evident that the sample obtained in the PROFIT survey is a sample of highly educated individuals, with one third of the sample being above the Masters level, almost 60% being at Masters level and above, and 86.4% being educated above the high school level. English respondents are not significantly differently educated on average, i.e., 16.9 years of education on average for the English respondents, compared to 16.8 for non-English respondents. Among the user groups, education is higher among parents of children (UG4), potential

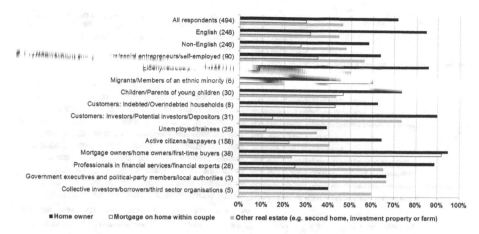

Fig. 4. Distributions by property ownership. *Notes*: The comments in Fig. 1 regarding the presentation format apply. Property ownership stems from Questions 11 and 12 in the questionnaire.

investors (UG6), active citizens (UG8), mortgage owners (UG9) and professionals in the financial services (UG10). It is notably lower among the unemployed and the elderly. It is also notably low among migrants (UG3), over indebted household members (UG5), but it is worth reminding that the numbers of observations are low for these groups.

Figure 4 presents the figures for property ownership stemming from questions 11 and 12 in the questionnaire of Annex I. More than 70% of the respondents are home owners, with some 30% having a mortgage. A figure close to 48% of the respondents own additional real estate apart from their home. It is also the case that English language respondents are significantly more likely to be home owners. Some 85% are home owners, compared to some 59% of non-English-language respondents. English language respondents are also more likely to have a mortgage. However, with respect to ownership of other real estate, differences between English and other language respondents are the opposite. Non-English respondents have higher figures for ownership of other real estate, but these differences from their respective counterparts are not statistically significant. The groups more likely to own property are the elderly, parents of children, investors, active citizens, mortgage owners and professionals in the financial services.

3.2 Financial Behavior

Our second set of characteristics of interest is related to financial behavior. We identify 9 sets of characteristics of interest and discuss them in the next paragraphs, namely: (i) savings, (ii) saving for retirement, (iii) keeping expense records, (iv) type of financial-advice receivership, (v) type of financial advisor, (vi) willingness to take financial risks, (vii) comparing financial service providers, (viii) financial experience

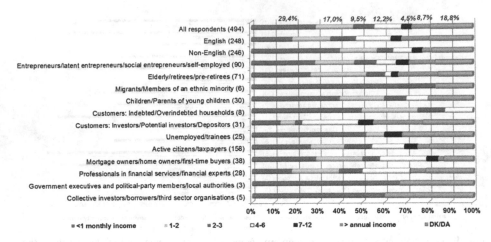

Fig. 5. The savings distribution. *Notes*: The comments in Fig. 1 regarding the presentation format apply. The amount of savings is grouped into the seven categories above: (i) < 1 monthly income, (ii) approximately 1–2 times the monthly income, (iii) approximately 2–3 times the monthly income (iv) approximately 4–6 times the monthly income, (v) approximately 7–12 times the monthly income, (vi) > annual income, and (vii) don't know or refuse to answer (merged from two distinct categories) Savings stem from Question 5 in the questionnaire.

and awareness of protection agency, and (ix) type of unsatisfactory financial experience. These will be discussed in short in this section.

Figure 5 presents the distribution of the amount of available savings as a fraction of income. This stems from question 5 of the questionnaire in Annex I. It is shown that 29.4% of the respondents have savings equivalent to an amount that is less than 1 monthly income. 17% of the respondents have savings equal to 1–2 times the monthly income, and an additional 9.5% have savings equivalent to 2–3 times that income. 12.2% and 4.5% have savings between 4–6 months of income and between 7–12 times monthly income, respectively. A figure equal to 8.7% gave savings greater than an annual income. It is worth noting that 18.8% of the respondents answered that they do not know or that they do not wish to answer. It is worth noting that individuals are split among these latter two categories almost equally 9.7% responded that they do not know, and 9.1% that they do not wish to answer. English language respondents. have higher savings on average. Among the user groups, those with higher savings appear to be investors, the active citizens, home owners/buyers, and the professionals in the financial services. The low overall figures resemble the globally low figures reported in the literature (e.g. [31]).

Figure 6 examines saving for retirement, in terms of ownership of a private retirement or insurance account, and the mental accounting regarding figuring out how much one needs to save for retirement. A figure close to 38% of individuals have private retirement or insurance account, with some 31% having figured out how much they need to save for retirement. The figures are remarkably close to the average figure of 35% for high-income OECD counties (the average global figure is 25%, reported by [8]). English language respondents perform significantly higher in both categories,

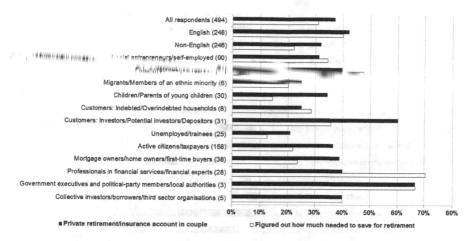

Fig. 6. Saving for retirement and retirement/insurance accounts. *Notes*: The comments in Fig. 1 regarding the presentation format apply. The figure stems from Questions 6 and 9 in the questionnaire.

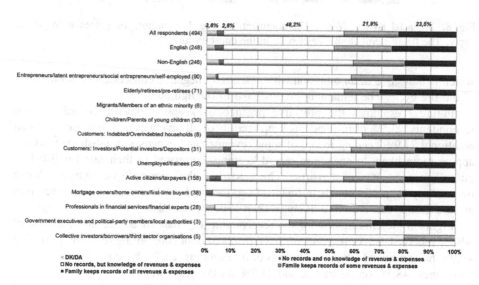

Fig. 7. Keeping expense records. *Notes*: The comments in Fig. 1 regarding the presentation format apply. The figure stems from Question 22 in the questionnaire.

compared to non-English language respondents. Potential investors are the one group that is much more likely to own private retirement and insurance accounts, with a figure close to 60%. Professionals in financial services and the elderly also have high figures, closer to 40%. 70% of the professionals in financial services know how much they need to save for retirement. Less than 50% of the elderly have done this accounting exercise,

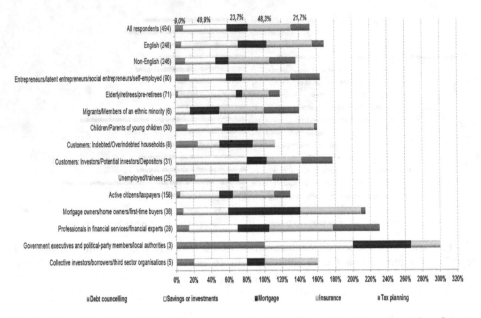

Fig. 8. Financial advice. *Notes*: The comments in Fig. 1 regarding the presentation format apply. The figure stems from Question 10 in the questionnaire.

which is indicative of an omission with potentially negative consequences at this stage of the life cycle. The remaining groups have lower figures.

Figure 7 examines the proactive practice of keeping expense records among respondents. 23.5% of respondents keep records of all revenues and expenses, with an additional 21.9% keeping record of some revenues and expenses. English respondents appear more likely to engage in record keeping, compared to their non-English language respondent counterparts. The elderly, the unemployed/trainees, home owners/buyers, professionals in financial services and the few government executives are the groups that are more likely to keep expense records.

Figure 8 presents figures for the receivership of financial advice and the type of advice received. 9% of respondents have received debt counselling recently, a striking 49.9% have received advice on savings or investments, 23.7% have received advice on mortgages, 48.3% on insurance, and 21.7% on tax planning.

Hence, savings/investments and insurance appear to be the most frequent financial counselling topics. English language respondents are more likely to have received advice overall and in these same three categories. They are less likely to have received debt counselling and tax planning advice. The groups more likely to have sought and received financial advising, presented in terms of the overall likelihood, are: professionals in financial services, owners/buyers/mortgage, investors, entrepreneurs, and parents. Active citizens and the elderly have lower figures overall. A very high figure is shown for government executives and political party members, reminding that there are only 3 respondents in this category.

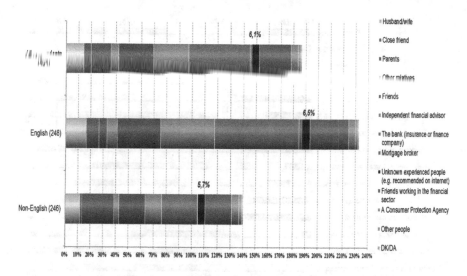

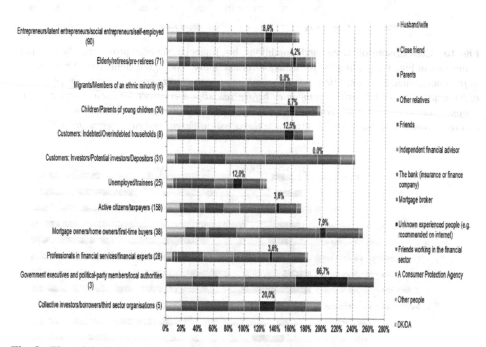

Fig. 9. Financial advisor. *Notes*: The comments in Fig. 1 regarding the presentation format apply. The figure stems from Question 14 in the questionnaire.

Figure 9 then presents the source of financial advice or financial advisor. The most frequent sources of financial advice are: the bank, insurance or finance company (47.8%), independent financial advisor (27.8%), friends (27.5%), and friends working in the financial sector (25.7%). There is an interesting figure of 6.1% for individuals

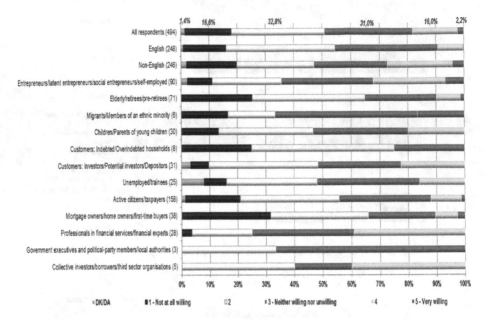

Fig. 10. Comparison of terms and conditions by financial service providers. *Notes*: The comments in Fig. 1 regarding the presentation format apply. The figure stems from Question 20 in the questionnaire stating: "*How often do you compare the terms and conditions for provision of financial services by various companies before you sign a contract for such a service? For instance, do you check loan terms, such as interest rate levels, maturity, collateral levels with different commercial banks or life insurance terms with different insurance companies?*"

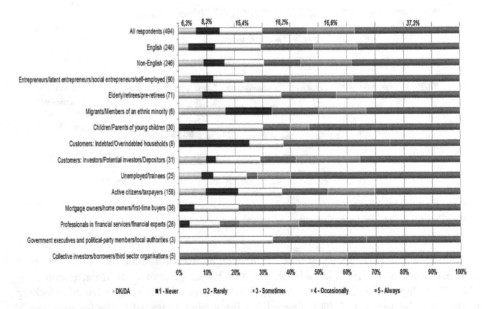

Fig. 11. Unsatisfactory financial-experience and awareness of financial consumer protection agency. *Notes*: The comments in Fig. 1 regarding the presentation format apply. The figure stems from Questions 19 and 23 in the questionnaire.

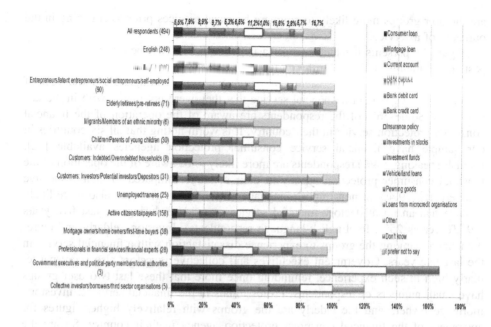

Fig. 12. Type of unsatisfactory financial experience. *Notes*: The comments in Fig. 1 regarding the presentation format apply. The figure stems from Question 24 in the questionnaire.

who are seeking for advice from unknown experienced people, e.g. people recommended on the internet. English language respondents appear much more likely to seek for advice overall and from both the formal and informal source aggregate categories.

The second panel of the figure presents the distributions of advice source by user group. Home/mortgage owners and potential investors are the two most likely user groups to seek for advice from all sources. Noting the small sub-sample size, over-indebted households appear likely to rely on informal sources of advice, and in terms of the informal/formal advice ratio, this is also the case with parents of children. Over indebted households also appear to seek for advice at the web (12.5%, compared to less than half that figure on average for all respondents). The government/party user group members also have high overall and web advice figures, but again its sub-sample is very small.

Figure 10 examines the habit of comparing financial services providers prior to a purchase. A fraction of 6.3% of the respondents did not respond to this question or did not know what to respond. 8.3% of the respondents never compare financial service providers, 15.4% rarely compare them prior to a purchase. 16.2% sometimes compare providers and 16.6% occasionally compare them. A third of the sample, i.e. 37.3% of the respondents, always compares financial service providers prior to a purchase of a service. The inspection of the figures for the user groups suggests that English-language respondents are less likely to be at the top categories for comparing financial service providers, compared to respondents to the questionnaire in other languages. Entrepreneurs, investors, unemployed, professionals and mortgage owners are the user groups more likely to compare financial services prior to engaging in the purchase of a service.

Figure 11 presents the incidence of a negative financial experience in the recent past among the respondents. Furthermore, it presents the prevalence of such experience alongside awareness of the existence of a financial protection agency. Some 32% of respondents have bought a financial service that they were not happy with in the last five years. Some 36% of the respondents are aware of the operation of the financial consumer protection service in their country. It is worth noting that all six countries in our sample have financial service consumer protection agencies available [32]. English-speaking (44%) respondents are more likely to be aware of the operation of the financial consumer protection agency in their country, compared to their respective counterparts (29.4% of non-English-language respondents). They are also more likely to have had an unsatisfactory financial service experience during the last five years (39.2% versus 25%, for English-language respondents). Investors, elderly and entrepreneurs are among the groups to experience dissatisfaction with a financial service in the last five years. Government executives and collective investors also score particularly high in such experience, reminding once more that these last two user groups have small numbers of respondents. Professionals in the financial services, investors, mortgage owners and the elderly are the groups with relatively higher figures for awareness of the financial consumer protection agency in their country. So are the migrants, noting again the few responses in this category.

Figure 12 elaborates on the previous question by presenting the type of unsatisfactory financial experience individuals have had in the last 5 years. 16.7% of the respondents prefer not to say what that experience was. 15.8% had a negative experience with investment funds and 11.2% with insurance policies. 9.7% had a problem with a bank deposit and 8.9% with a current account. English-language respondents are much less likely not to say what their negative experience was, compared to non-English speaking counterparts. User groups 11 and 12 of government/party executives and collective investors/borrowers/third-party organizations appear to have the highest incidence overall, with high figures for relatively standard financial transactions, e.g., consumer loans, bank and current accounts, and insurance policies. Given the small numbers of observations for these two groups, this observation is not elaborated upon any further. Other patterns worth highlighting among user groups involve the high figure of a negative experience with mortgages from mortgage/home owners. Moreover, there are high figures of negative experiences with investment funds among the investors and the elderly. Entrepreneurs, investors and mortgage owners are also likely to express relatively high figures of an unsatisfactory experience with insurance policies.

3.3 Financial Knowledge

This sub-section turns the reader's attention to respondents' self-assessment of their financial knowledge and the typical sources of information. The inspection of the Fig. 13 indicates that 4.1% of the respondents find their financial knowledge is very low and 16.8% find it to be relatively low. Nearly half of the respondents, i.e. 46.6%,

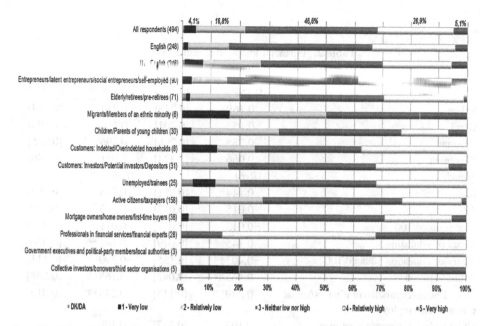

Fig. 13. Financial knowledge self-assessment. *Notes*: The comments in Fig. 1 regarding the presentation format apply. The figure stems from Question 13 in the questionnaire.

assess their financial knowledge as neither high nor low. 26.9% find it to be relatively high and an additional 5.1% assess they have very high financial knowledge. The observations on the bulk of respondents in the middle category and the very low figure at the top category are interpreted as clearly indicative of the existence of a market for personal financial training. English-language respondents are somewhat less likely to be at the bottom two categories of financial knowledge. However, it is worth noting that the differences are non-existent for the top four knowledge categories for both groups.

Expectedly, professionals in the financial services are the most confident for their financial knowledge, with some 86% being at the top two categories, and none in the bottom two. They are followed by entrepreneurs and investors. This is again illuminating in highlighting that it is these three groups that should be considered as the most advanced or sophisticated connoisseurs of finance. The literature review in Sect. 2 can also justify this assertion (e.g., [17, 30, 35], inter alia). Higher occurrence in the bottom categories appear among parents of young children, migrants, over-indebted households and the unemployed. Again, the findings on the four user groups that have limited observations cannot be elaborated upon much further.

In order to shed some more light in the ranking of the groups in terms of their self-assessed financial knowledge, Table 2 and Fig. 14 present results from regression estimates of financial knowledge.

Linear regressions are presented in Table 2 and marginal effects from ordered probit regressions for the five categories of financial knowledge are presented in Fig. 14. Column 1 of Table 2 presents estimates from specifications including language (6 categories), 2nd order polynomials in age and education and attitude to financial

Table 2. Financial knowledge regressions

	(1)		(2)	
English	0.358**	[0.141]	0.356***	[0.026]
Italian	0.169	[0.243]	0.214	[0.144]
Greek	0.153	[0.201]	0.152	[0.198]
Croatian	0.285	[0.208]	0.296***	[0.065]
Slovenian	0.066	[0.224]	0.051	[0.049]
French	[*Ref.*]		[*Ref.*]	
Male	0.144*	[0.081]	0.146	[0.155]
Age	0.015	[0.017]	0.013	[0.009]
Age squared/100	−0.013	[0.017]	−0.011	[0.012]
Years of education	0.145***	[0.049]	0.141**	[0.036]
Years of education squared/100	−0.403***	[0.155]	−0.390**	[0.111]
Attitude to risk	0.165***	[0.042]	0.168**	[0.058]
UG$_1$: Entrepreneurs/latent e'neurs/social e'neurs/self-employed	0.249**	[0.116]	0.245***	[0.013]
UG$_2$: Elderly/retirees/pre-retirees	0.107	[0.147]	0.109	[0.062]
UG$_3$: Migrants/Members of an ethnic minority	–		−0.328	[0.698]
UG$_4$: Children/Parents of young children	−0.071	[0.179]	−0.073	[0.065]
UG$_5$: Customers: Indebted/Overindebted households	–		0.195	[0.426]
UG$_6$: Customers: Investors/Potential investors/Depositors	0.167	[0.171]	0.166*	[0.070]
UG$_7$: Unemployed/trainees	0.278	[0.199]	0.275	[0.137]
UG$_8$: Active citizens/taxpayers	[*Ref.*]		[*Ref.*]	
UG$_9$: Mortgage owners/home owners/first-time buyers	0.138	[0.163]	0.133	[0.069]
UG$_{10}$: Professionals in financial services	1.053***	[0.142]	1.044***	[0.057]
UG$_{11}$: Government exec's, political-party members/local authorities	–		0.218	[0.249]
UG$_{12}$: Collective investors/borrowers/third sector organisations	–		−0.415	[0.315]
UG$_5$, UG$_{11}$, UG$_{12}$	−0.068	[0.239]	–	
Constant	0.535	[0.560]	0.593	[0.358]
Linear prediction	3.1197		3.1197	
No. of Observations	493		493	
R^2	0.191		0.195	

Notes: Estimates are from linear regressions. Asterisks denote the following levels of significance: * p < 0.10, ** p < 0.05, *** p < 0.01

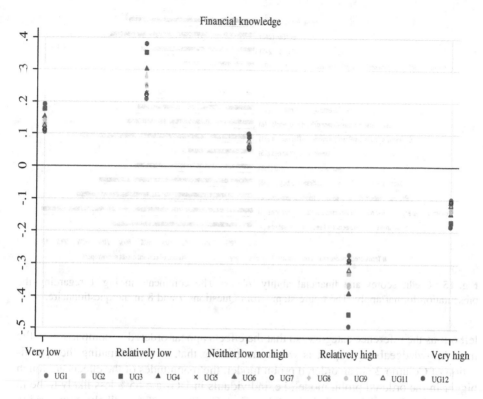

Fig. 14. Financial knowledge: marginal effects from ordinal regressions. *Notes*: The figure presents marginal effects from ordered probit regressions. The reference category is UG10: Professional in financial services. (Color figure online)

risk. Finally, dummy variables for the 8 user groups that received adequate numbers of respondents are included, along with a dummy variable for all 4 groups that did not receive many responses, i.e. UG_3, UG_5, UG_{11}, and UG_{12}. Then, the specification of Column 2 also allows for these four user group categories to be included separately. The reference category among user groups is UG_8, i.e. the broader group of active citizens. The estimates presented confirm that professionals in financial services are some 30 percent more self-confident regarding their financial ability, compared to active citizens and based on the liner predictions of the models. Individuals in user group 1, i.e. entrepreneurs and related, are some 8% more confident regarding their knowledge, compared to active citizens. Then, in Column 2, individuals in user group 6, i.e. investors and related, are some 5.3 percent more confident compared to active citizens.

The marginal effects from ordered probit models, which are plotted in different colors at Fig. 14, confirm that the three groups identified above appear as the most confident regarding financial knowledge. The figure plots the marginal effects from a specification similar to that of Column 2 of Table 2, with the difference that UG_{10} is

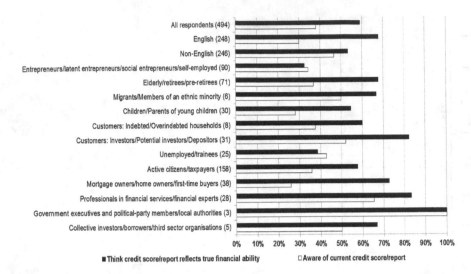

Fig. 15. Credit scores and financial ability. *Notes*: The comments in Fig. 1 regarding the presentation format apply. The figure stems from Questions 7 and 8 in the questionnaire.

left our as the reference category, so that the effects appear ordered in comparison to the most knowledgeable group. It is worth mentioning that, when estimating the specification of Column 2 via an ordered probit model, the magnitudes of the effects are much bigger in the ordered probit model, i.e. individuals in UG_1 are 65% less likely to be in category 1 of financial knowledge and those in UG_6, some 40% less likely, compared to active citizens (UG_8). In contrast, they are some 50% and 30% more likely to be in category 5 of financial knowledge, respectively, compared to active citizens. Professionals of the UG_{10} are always ranking on top in terms of confidence in financial knowledge (this group is not shown in Fig. 14, as it is the reference group).

Then, Fig. 15 stems from an inquiry into the relationship between self-assessed and institutionally-assessed financial capability. The figure presents the fractions of individuals who are aware of their current credit score/report, side-by-side with the fraction of individuals who think their credit score/report reflects their true financial ability. Overall, less than 40% of respondent are aware of their current credit score/report, while some 60% of the respondents do not think their credit score/report reflects their true financial ability.

3.4   Financial Awareness: News, Information Retrieval and Financial Sentiment

This sub-section shifts the attention to the question that is related to the willingness to take financial risks, familiarity with financial trends and sources of information with respect to financial trends. Figure 16 examines the expressed willingness to take financial risks. A fraction of 16.6% of the respondents is not at all willing to take financial risks, while 32.8% are not too willing. A fraction of 31% is neither willing nor unwilling to take financial risks. 16% of the respondents are somewhat willing and a reminder 2.2% is very willing to take financial risks. English-speaking respondents appear less willing to be at the top categories of the willingness to take financial risks and are less willing on average. Among user types, professionals in the financial services, entrepreneurs and investors are more willing to take financial risks. It is worth noting that these findings accord well with the battery of the empirical evidence in financial economics [35, 36]. A point of caution from the literature is that for professionals in financial services, the willingness to take risks appears to be counter-cyclical and affected a lot by fear. There is a relatively high figure for collective investors, but it is worth reminding that the number of observations received was small for this category. It is worth noting that the observation on willingness to take risks alongside self-assessed financial knowledge is an important empirical observation that directly proposes a rearrangement for the previous ad hoc classification of user groups, based on the level of experience. It is observed and proposed that the above groups should be considered as the experienced users, i.e. entrepreneurs, investors and professionals in financial services, instead of user groups 10–12.

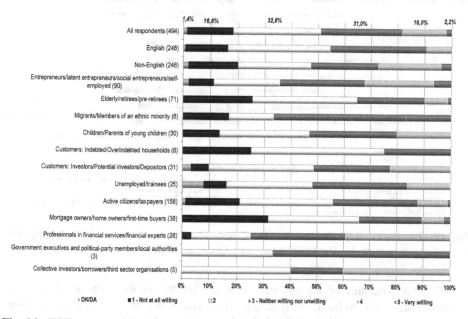

Fig. 16. Willingness to take financial risks. *Notes*: The comments in Fig. 1 regarding the presentation format apply. The figure stems from Question 15 in the questionnaire.

The analysis of the answers of Fig. 16, concerning the willingness of the participants to take financial risks, revealed that almost 50% of the participants are risk averse which means that are totally unwilling to take risks while only 18% are risk lovers. The remaining 32% is defined as risk neutral. The results are in line with the findings of relative experimental research, which tend to the notion that humans are mostly risk averse.

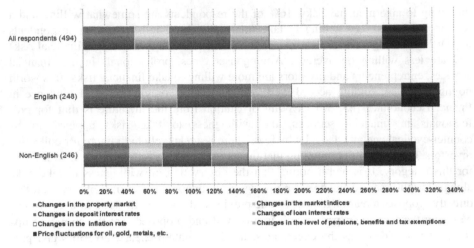

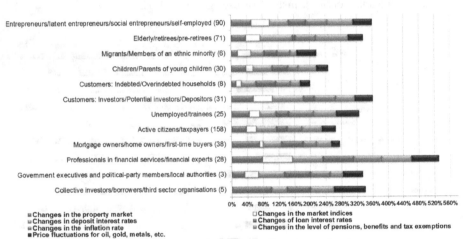

Fig. 17. Familiarity with financial trends. *Notes*: The comments in Fig. 1 regarding the presentation format apply. The figure stems from Question 16 in the questionnaire.

The functionalities of the platform that are related to the forecasting will be developed according to the above findings. More specifically, the platform will provide predictions for the direction of Stock Markets, Commodity Markets and Exchange Rates rather than point forecasts. These predictions, at least theoretically, are crucial for asset allocation decisions between financial assets and risk free interest rate investments. The investment decisions based on the direction of financial assets are most of the times related to relatively lower returns and risks, compared to more risky investment decisions based on the future realizations of the price of a financial asset.

Figure 17 presents an inquiry into the specifics of financial information, presenting the cumulative distributions of expressions of familiarity with specific financial trends. The lowest figure overall (31.6%) appears to be that for changes in market indices.

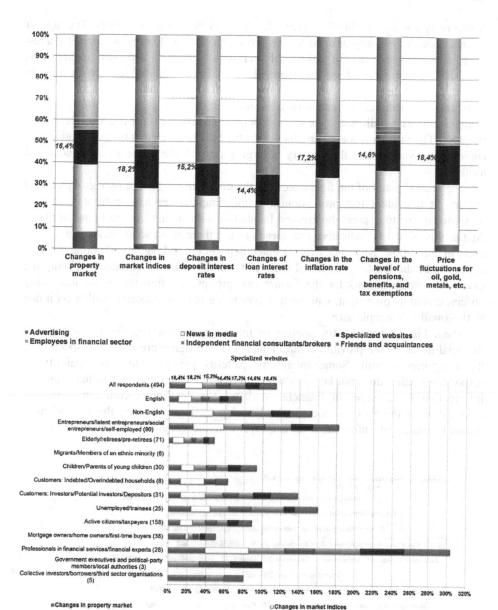

Fig. 18. Sources of financial information and the use of specialised websites. *Notes*: Panel A presents the sources of financial information for all respondents and Panel B presents the use of specialised websites for financial information by user group. The comments in Fig. 1 regarding the presentation format apply for Panel B. The figure stems from Question 17 in the questionnaire.

A relatively low figure (34.9%) appears for changes in loan interest rates. The highest figures, around 57%, appear to be for the two categories on changes on deposit interest rates and changes in the levels of pensions, benefits and tax exemptions. English-language respondents are more likely to have high figures for changes in deposit interest rates. This indicates that the incidence of information as deposit interest rates and tax, benefits and exemptions is high for this group, while information on changes in market indices and the macroeconomic environment for interest rates is less frequent. The intension was to emphasize on the two latter elements. Information that entails large variation at the country is largely beyond the scope of an educational toolkit.

The second panel of the figure presents the distributions for the user groups in our inquiry. It is evident that professionals in the financial services, entrepreneurs and investors appear to express the greatest familiarity with changes in the financial plane. Again, this is clearly indicative that these are the three most sophisticated groups of users in our sample. Migrants, over-indebted households and parents of young children appear to be the least sophisticated users in terms of financial awareness. Noting the caveat of small numbers for the former two groups, we find this is an interesting observation at its own right, without however being able to elaborate further on it due to the small sub-sample size.

Then, Fig. 18 presents the sources of information regarding the 7 categories of financial trends in the previous figure. Panel A of the figure presents an inspection of the categories overall. Some interesting patterns prevail. The vast majority of respondents either does not know or does not answer, when it comes to the source of information for changes in financial markets. The first most common source of information is the news in the media. For deposit interest rates, the second most common source of information is employees in the financial sector. For the vast

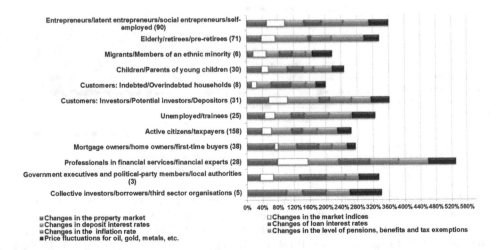

Fig. 19. Sources of information when choosing a company to buy a financial service from. *Notes*: The comments in Fig. 1 regarding the presentation format apply. The figure stems from Question 21 in the questionnaire.

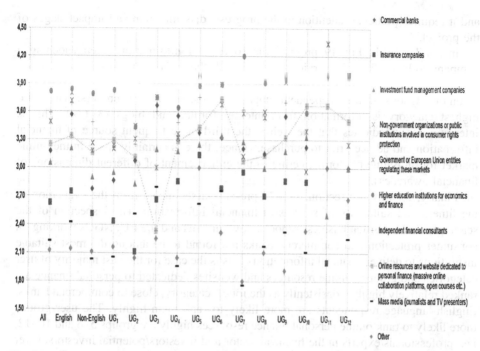

Fig. 20. Evaluation of the fitness and suitability of sources of financial information. *Notes*: The figures stem from evaluations on a scale ranging from 1 (least suitable) to 5 (most suitable). The figure stems from Question 18 in the questionnaire.

majority of categories, the second most common source of information stems from specialized websites, with figures consistently between 15–18% for all categories of changes in the financial markets. The discrepancy between the high figures for those who do not know how to seek for information regarding changes in financial trends and the significant but still low figure for the use of specialized websites again indicates that the PROFIT project is likely to cater to a strong existing need in the market for financial knowledge and information.

The second panel of the figure emphasizes on the use of specialized websites and presents its usage per trend and by user category. It is observed that English-language respondents are less likely to use specialized websites for all categories of financial trends, compared to their other language counterparts. Professionals in financial services, entrepreneurs and investors are again the three most likely groups to use specialized websites. Again this indicates these are the most sophisticated user types in terms of financial awareness of means and resources. An interesting pattern is that the unemployed are also a group that appears more likely to use specialized websites as sources of information for financial trends. It is worth noting that among the 6 migrants in our sample none reported using specialized websites for financial information. This is also indicative of another potential major challenge, i.e. the means of digital inclusion for financially excluded populations. This is a challenging endeavor globally, and it requires particular attention in the progress, dissemination and impact stages of the project.

In similar spirit, Fig. 19 presents the sources of information when choosing a company to buy a financial service from. Overall, the highest category is for informal sources. 44% of the sample receives advice from friends and relative. 38.7% receive recommendations from independent financial consultancies or brokers. The third highest category is information from internet forums and blogs (31.6%). This is interesting as it indicates that the web is the third more frequent source of financial information and advice prior to purchase. Hence, there is certainly space in the online market for an online platform catering to the enhancement of different dimensions of financial awareness.

Finally, for this sub-section, Fig. 20 presents assessments from the evaluation of the fitness and suitability of sources of financial information. The inspection of the scatterplot indicates that business schools, governments and the EU, NGOs working on consumer protection and commercial banks are found to be among the most suitable institutions to deliver financial information. This is the case for the vast majority of user categories and groups. Online resources and websites dedicated to personal finance are ranked relatively highly, consistently as the fourth category, close to commercial banks. English-language respondents are more likely to rank them highly. The user groups more likely to rank online personal finance resources highly are groups 5–6 and 10–12, i.e., professionals/experts in the financial sector and investors/potential investors. Over indebted household members, government and party executives and collective investors are also likely to rank them really highly. However, we must again note the small numbers of responses in these latter three groups.

3.5   Financial Capability: Generic Financial Knowledge and Specific Financial Literacy Training

The next section of the questionnaire pointed the respondents' attention to generic aspects of financial knowledge that could be of relevance and to specific financial literacy aspects that stem from the literature and the current practice. Figure 21 presents respondents' evaluation of fourteen aspects of financial knowledge that could be seen as relevant. These are more related to generic financial capability, which is linked to the experience of usage of financial products. Question 25 was asking: *"Which of the following aspects do you believe should be a required part of a financial training course for citizens?"* Panel A of the figure presents bars for the overall evaluation. Then, Panel B of the figure presents a scatterplot highlighting preferences by group.

The general observation is that all fourteen types of information receive a high ranking, i.e., above two thirds of the respondents think that this information is important. The aspects that rank the highest, i.e. above 90%, are: *"What information should a user pay attention to when signing a contract with a bank or another financial institution"*, *"How does the pension scheme work and what methods are available to secure one's old age income"*, *"How to establish one's own financial targets and formulate a personal current financial plan"*, *"What consumer protection laws are*

Fig. 21. Aspects of relevance in financial training. *Notes*: The comments in Fig. 1 regarding the presentation format apply. The figure stems from Question 25 in the questionnaire.

available and what one needs to do when one's consumer rights are violated", *"How to do personal management so as not to get over-indebted ("go in red")"*.

The second group of categories, scoring between 80–90% are: *"How to plan purchases of durables (car, apartment) and evaluate one's abilities to implement them"*, *"Banking services – current accounts, saving deposits and credit cards"*, *"Sources of information on financial services, interpretation of the information and ways to differentiate advertising from objective information"*, and *"Insurance and related products"*. Moreover, the third category of interests, scoring between 70–80% involves: *"Mortgage loans"*, *"What parameters are used to compare the services offered by banks and other financial companies"*, *"Understanding public finances, e.g. basic macroeconomic national and international facts and economic policies"*, *"Planning for starting one's own business"*. Finally, there is one category, which around 68% of the respondents see as relevant to financial training, namely *"Capital*

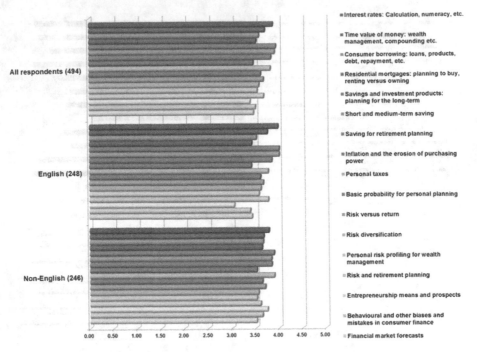

Fig. 22. Specific financial literacy aspects. *Notes*: The comments in Fig. 1 regarding the presentation format apply. The figure stems from Question 32 in the questionnaire.

markets, stock and investment funds". This is perhaps indicative of the more advanced nature of this type of information compared to the basic financial training.

The second panel of the figure presents the scatterplot by user group status and also by English-speaking questionnaire response status. The ranking of the categories is more or less confirmed. Although it seems that there are minor differences across user groups in terms of the relative ranking of categories, there are no major discrepancies when it comes to the categories that arise at the top. The legend of the figure presents these categories in order of overall importance, as analyzed in the previous paragraph. Some observed discrepancies seem to arise mostly from the four groups that did not receive many responses, i.e., UG_3, UG_5, UG_{11}, and UG_{12}. A final interesting observation stems from inspecting the preferences of the more advanced groups in the categorization. For UG_1 of entrepreneurs and related, the top categories are very close to each other, in figures well above 90%. For UG_6 of investors and related, understanding public finances rank really high at some 87%, along with understanding banking services. Finally, for UG_{10}: professionals in financial services, understanding insurance and related products scores close to the top categories at 89%. Understanding mortgage loans and the bottom category for understanding capital markets, stocks and investments, along with sources of information and interpretation for financial services score very highly, i.e. above 80%. These are features that are different from the remaining groups.

Then, Figs. 22 and 23 present a thorough investigation into the specific financial-literacy content of a financial education toolkit, which respondents see as relevant. Question 32 was asking, "Rating from 1 to 5, *how useful would you find an online platform that provides financial literacy/basic financial education training on* ⁿ... Seventeen categories were offered to the respondents, stemming from best current practices in personal finance and recent developments in the financial literacy literature (e.g. [2, 5, 11, 18, 24, 37–39]. These are: (1) Interest rates: Calculation, numeracy, etc.; (2) Time value of money: wealth management, compounding etc.; (3) Consumer borrowing: loans, products, debt, repayment, etc.; (4) Residential mortgages: planning to buy, renting versus owning; (5) Savings and investment products: planning for the long-term; (6) Short and medium-term saving; (7) Saving for retirement planning; (8) Inflation and the erosion of purchasing power; (9) Personal taxes; (10) Basic probability for personal planning; (11) Risk versus return; (12) Risk diversification; (13) Personal risk profiling for wealth management; (14) Risk and retirement planning; (15) Entrepreneurship means and prospects; (16) Behavioral and other biases and mistakes in consumer finance; (17) Financial market forecasts.

It must be noted that more than half of respondents agreed that all 17 categories are relevant. The inspection of the bars in Fig. 22 suggests that the highest rankings, on average from 1 to 5, are obtained for: (5) Savings and investment products: planning for the long-term; (6) Short and medium-term saving; (7) Saving for retirement planning; (1) Interest rates: Calculation, numeracy, etc.; (9) Personal taxes; (2) Time value of money: wealth management, compounding etc.; and (14) Risk and retirement planning. Interestingly, the lowest ranking is obtained for: (15) Entrepreneurship means and prospects; (16) Behavioral and other biases and mistakes in consumer finance; (17) Financial market forecasts; (8) Inflation and the erosion of purchasing power; and (13) Personal risk profiling for wealth management. With the exception of inflation, the remaining categories are considered potential PROFIT novelties into the financial-literacy literature and practice. An interesting pattern emerges from the comparison between English-speaking respondents and non-English speaking respondents. English-speaking respondents are more likely to give a lower average ranking to the last four categories, with the exception of inflation and compared to non-English speaking respondents.

Finally, Fig. 23 provides a thorough investigation of the responses regarding the specific financial-literacy training categories. The first panel of the figure presents a scatterplot for the categories of responses by user group. Then, the second panel presents bars for cumulative figures for the responses in the top two categories of the 5-point Likert scale. An interesting pattern revealed in the scatterplot of the first panel is that among the 4 basic concepts of financial-literacy, namely (i) interest rates/calculation, numeracy; (ii) time value of money: compounding, wealth management etc.; (iii) inflation and the erosion of purchasing power; and (iv) risk diversification (e.g. [40]), it is mere numeracy in terms of interest rate calculation that obtains the highest ranking among respondents. The time value of money/compounding receives a relatively high ranking, with the remaining basic concept categories ranking in the middle among most groups. Some interesting exceptions arise among the expert groups, i.e. risk diversification ranks on top for UG6, i.e. investors and related. Entrepreneurship means and prospects ranks on top for UG1, i.e. entrepreneurs and

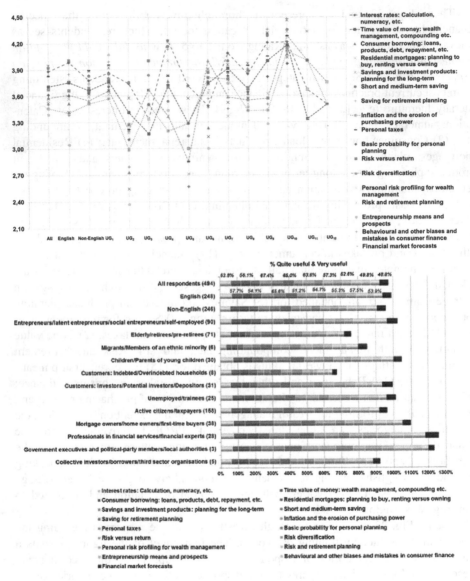

Fig. 23. Specific financial literacy aspects by group. *Notes*: The comments of Fig. 22 apply. The second panel of the figure presents cumulative figures for the top 2 categories in the 5-point Likert scale.

related. For the remaining groups, this ranks particularly low and close to the bottom. A final interesting pattern emerges from the observation of the ranking for the top experts, i.e. UG10 of professionals in financial services. For that group all proposed categories rank high, i.e. with an average of above 3.9 out of 5. Entrepreneurship means and prospects ranks below 3.9 and at the bottom for that group. One interpretation of this high and very near ranking of our proposed categories among the experts in this group is that it confirms and validates the importance of the material proposed, including the novelties in that material, e.g. on public finance, entrepreneurship, behavioral finance etc.

The relevance and validity of the 17 categories of proposed material is also confirmed by the cumulative figures for the top 2 categories in the 5-point Likert scale, which are presented in the second panel of Fig. 23. There, it is evident that professionals in financial services give a consistently high ranking in the vast majority of categories, with the only exception of entrepreneurship means and prospects. That category is also one of the two, which less than 50% of respondents overall in the sample find as important. Specifically, 49.8% have replied that they find entrepreneurship means and prospects as important. Then, the second lowest category is for financial market forecasts, which 48.8% of the respondents saw as quite useful or very useful. This is a category that is not planned as basic financial-literacy material, but is relevant to the financial awareness sections of the platform.

Other interesting observations involve the high ranking of the vast majority of the response categories by UG_{11}, i.e. government executives, party members and related, noting that only 3 respondents provided answers in that group. UG_9, i.e. mortgage and home owners/buyers also give a consistently high ranking to the vast majority of the 17 categories. Other groups that provide high cumulative rankings are the parents of young children in UG_4 and entrepreneurs and related in UG_1. The indebted households in UG_5 and the elderly in UG_2 provide the lowest cumulative rankings. These groups are more likely to have specific thematic interests, e.g. retirement planning or time value of money, etc.

Table 3. Summary statistics for technological literacy question

	No. of observations		Median	Mode
	Valid	Missing		
Communication	493	1	5	5
Searching the Internet for information	492	2	5	5
Online communication	493	1	5	5
Entertainment	492	2	4	5
Navigation	491	3	3	3
E-banking services	492	2	3	3
Shopping online	492	2	2	2
Get local updates	491	3	4	5
Altruistic causes	490	4	2	2
Participate in online social platforms	491	3	4	5

Notes: The table stems from textual analysis of the responses to Question 26 in the questionnaire.

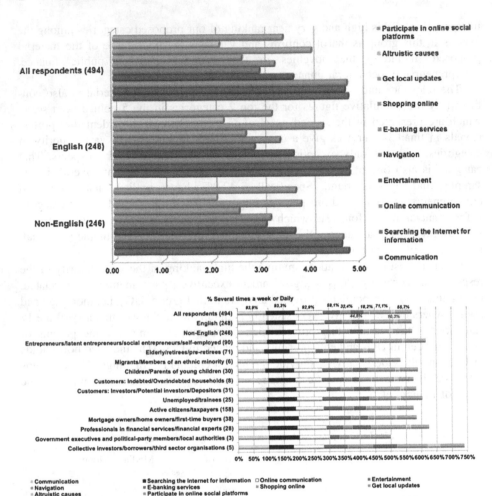

Fig. 24. Technological literacy. *Notes*: The figures stem from evaluations on a scale from 1 to 5 (1 = never, 2 = once a month or less, 3 = several times a month, 4 = several times a week, 5 = daily). The figure stems from Question 26 in the questionnaire.

3.6 Technological Literacy and Incentives

The following questions were intended to: (a) assess participants' technological literacy, (b) investigate the participants' attitudes towards such a platform, and (c) help us identify the incentives that appeal to different target users so we would incorporate more effective incentive mechanisms into the web platform.

Figure 24 presents facets of technological literacy, stemming from responses on a 1–5 Likert scale regarding the frequency with which one uses their computer, mobile phone or tablet for the following 10 reasons. Each question is a 5-point Likert scale, including a number of 5-point Likert items. The response categories were: 1 = never, 2 = once a month or less, 3 = several times a month, 4 = several times a week,

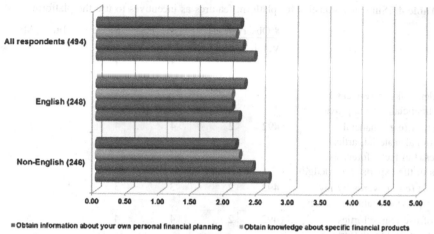

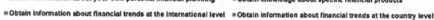

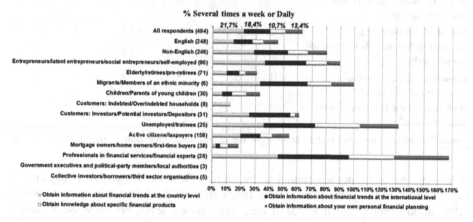

Fig. 25. Using the internet for financial information. *Notes*: The figures stem from evaluations on a scale from 1 to 5 (1 = never, 2 = once a month or less, 3 = several times a month, 4 = several times a week, 5 = daily). The figure stems from Question 27 in the questionnaire.

5 = daily. As can be seen in Fig. 24, as well as in the Table 3, the participants were familiar with these specific actions, since most of them perform them on a daily basis. They use navigation and e-banking less frequently (several times a month), and they shop online or contribute to altruistic causes online once a month or less. Entrepreneurs/latent entrepreneurs/social entrepreneurs/self-employed, professionals in financial services/financial experts, as well as collective investors/borrowers/third-sector organizations (respondents who can possibly be characterized as more experienced) score higher, as depicted in Fig. 24.

Figure 25 presents the frequency of Internet usage for the purposes of obtaining four different types of financial information. Again, responses were on a 1–5 Likert scale for the frequency with which one uses their computer, mobile phone or tablet for the following 10 reasons. The response categories were: 1 = never, 2 = once a month

Table 4. Summary statistics for platform features as incentives to use the platform

	# Observations		Median	Mode	Min	Max
	Valid	Missing				
Services (e.g. links to services by my current financial service provider, links to services by other financial service providers)	493	1	4	4	1	5
Resources (e.g. financial educational material, articles, quizzes, fan facts, forecasts, access to the experts' knowledge)	492	2	4	4	1	5
Rewards (e.g. rewards by my current provider, rating schemes)	492	2	3	3	1	5
Ease of use (e.g. effortless navigation on the platform, usability, user - friendly and attractive design, etc.)	492	2	4	4	1	5
Social interactions (e.g. discussion forums, chat, posts to social media)	492	2	3	3	1	5
Personalised Recommendations (e.g. for content of my preferences, for users of similar interests/profiles)	491	3	3	4	1	5

Notes: The table stems from textual analysis of the responses to Question 29 in the questionnaire.

or less, 3 = several times a month, 4 = several times a week, 5 = daily. As depicted in Fig. 25, most of the participants do not frequently use the Internet in order to obtain information about financial trends at local/international level, but they use it once a month or less in order to obtain info regarding financial products or their own personal financial planning, with non-English language respondents scoring higher regarding using the Internet for obtaining financial information at the country and international level. As it was deduced from the participants' answers in Question 14 of the online questionnaire (If you needed to know something about financial services, then who would you contact first?), most of the respondents usually prefer to seek financial advice from the bank (insurance or finance company), independent financial advisors, and friends, rather than a web platform.

Figure 26 and Table 4 presents responses on the platform features that are likely to act as incentives to use the platform on a regular basis. Responses were on a 1–5 Likert scale, with 1 representing "Not at all likely" and 5 representing "Very likely". These features included services, resources, rewards, ease of use, social interactions, and personalized recommendations. It was found that their attitude towards the afore-mentioned features was mostly positive (mode of responses 4). Responses were more neutral regarding *Rewards* and *Social interactions* (mode of responses 3), as can also be seen in the following Table. The platform feature which elicited the most positive

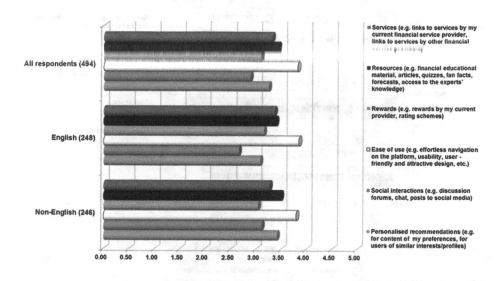

% Quite likely & Very likely

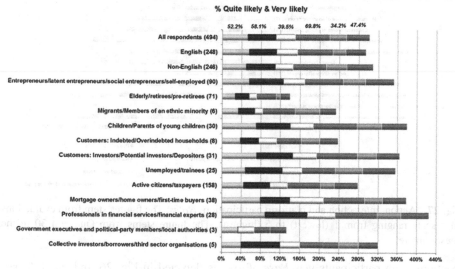

Fig. 26. Platform features as incentives to use the platform. *Notes*: The figures stem from evaluations on a scale ranging from 1 (not at all likely) to 5 (very likely). The figure stems from Question 29 in the questionnaire.

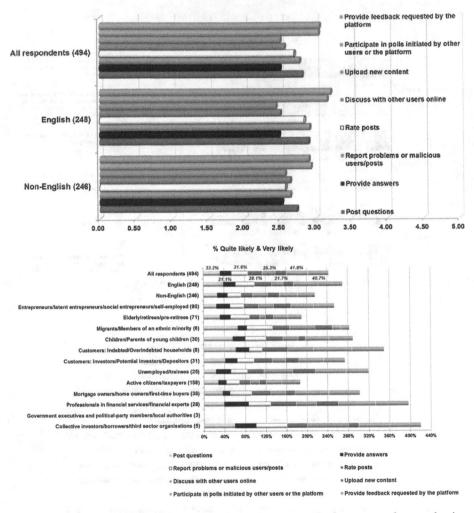

Fig. 27. Action, interaction and platform participation. *Notes*: The figures stem from evaluations on a scale ranging from 1 (lowest) to 5 (highest). The figure stems from Question 30 in the questionnaire.

reactions from participants was *Ease of use,* as depicted in Fig. 26. Indeed, 170 participants (34.4%) answered "Very likely" and 175 participants (35.4%) answered "Quite likely" (69.8%) regarding *Ease of Use*. Scores were also high regarding *Resources* and *Personalized Recommendations* (see Figs. 26 and 27).

Figure 27 presents the responses on the likelihood of performing 8 types of predetermined actions (e.g. post questions, provide answers, rate posts, etc.) when participating in an online platform. Once again, responses were on a 1-5 Likert scale, with 1 representing "Not at all likely" and 5 representing "Very likely". The participants were mostly neutral, as can be seen in Fig. 27. They were more positive regarding participating in polls and providing feedback to the platform, maybe because these actions do not require that much effort compared with creating & uploading their own

Table 5. Summary statistics for platform features as incentives to regularly contribute to the platform

	No. of observations		Median	Mode	Min	Max
	Valid	Missing				
Small monetary rewards (e.g. a tablet) according to my contributions to the platform	491	3	3	3	1	5
Social Status and Reputation (the level of my contributions to the platform can be highlighted in my profile)	491	3	3	3	1	5
Gamification Elements (e.g. leaderboards presenting the top contributors, awarded certain badges according to contributions, etc.)	491	3	2	1	1	5
Access to more advanced and/or moderation data and rights according to my contributions	491	3	3	4	1	5
Receive feedback and recognition from other users according to my contributions	491	3	3	3	1	5
Career opportunities (demonstrate my skills/expertise) in the platform according to my contributions	491	3	3	3	1	5
Social interactions and invitations to social events according to my contributions	491	3	3	4	1	5
More accurate personalised recommendations (e.g. for content of my preferences, for users of similar interests/profiles)	491	3	3	4	1	5

Notes: The table stems from textual analysis of the responses to Question 31 in the questionnaire.

material. Some differences were also identified between user categories. For example, more experienced user groups like professionals in financial services/financial experts and collective investors/borrowers/third sector organizations are more willing to provide answers, participate in polls and provide feedback to the platform. On the other hand, more basic user groups like migrants/members of an ethnic minority, parents of young children and customers: indebted/over indebted households are more willing to post questions on the platform, reflecting their need for gaining knowledge from the experts.

Figure 28 and Table 5 presents the responses on the importance of eight predetermined types of incentives that could induce regular contribution by the user to the platform. More specifically, these incentives were the following: (a) small monetary

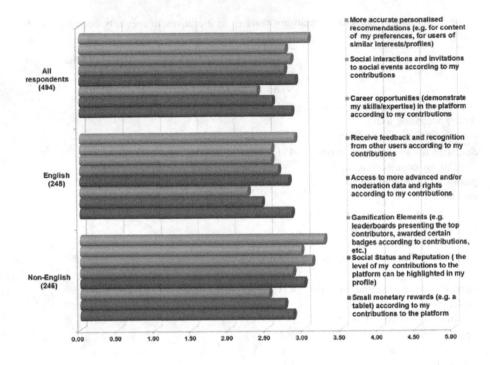

Fig. 28. Features that could act as incentives to contribute. *Notes*: The figures stem from evaluations on a scale ranging from 1 (not at all likely) to 5 (very likely). The figure stems from Question 31 in the questionnaire.

rewards, (b) social status and reputation, (c), gamification elements, (d) feedback and recognition from other users, (e) career opportunities, (f) social interactions, and (g) more accurate personalized recommendations. Once again, responses were on a 1–5

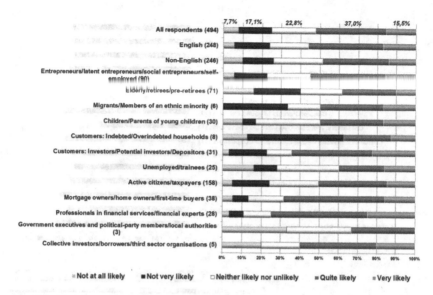

Fig. 29. Likelihood to use platform for updates on personal financial needs. *Notes*: The figures stem from evaluations on a scale ranging from 1 (not at all likely) to 5 (very likely). The figure stems from Question 28 in the questionnaire.

Likert scale, with 1 representing "Not at all likely" and 5 representing "Very likely". As can be seen, *Access to more advanced and/or moderation data and rights,* according to my contributions, *Social interactions and invitations to social events* according to my contributions, and *More accurate personalized recommendations*, elicited a higher number of positive responses. Unemployed/trainees have a more positive attitude to features that could act as incentives to contribute to the platform, with *Career opportunities* being the most prevalent incentive. *Receive feedback and recognition from users based on my contributions* is an important incentive for professionals in financial services/financial experts, together with having *access to more advanced data and moderation rights*, while *social status and reputation* is a more likely incentive to this group compared to the other ones. Parents of children and unemployed/trainees had a more positive attitude towards *gamification* compared to other user groups. Lastly, the most likely incentive for indebted/over indebted customers was *small monetary rewards*. It should be noticed that only elderly/retirees/pre-retirees seem to be less interested in incentives for using the platform, as it is depicted in Fig. 28.

As a final note, it should be mentioned that more advanced participants (as regards their financial status and needs) are more regularly informed regarding financial issues, they seem more positive in using such a platform and providing feedback and know-how to other users on the platform. In general, *access to more advanced data and/or moderation rights according to contributions* and *more accurate personalized recommendations* are the most prominent incentives to participate for all users.

Figure 29 presents responses on the likelihood to use an online platform to get updates on one's personal financial needs based on Question 28. Responses were on a 1–5 Likert scale, with 1 representing "Not at all likely" and 5 representing "Very

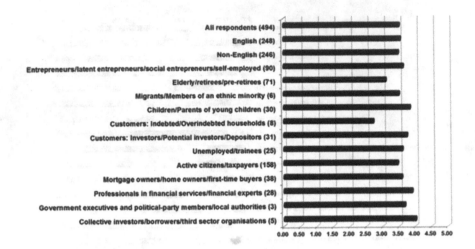

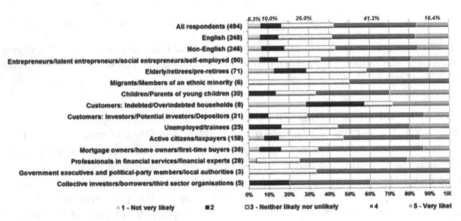

Fig. 30. Likelihood to use the platform for updates on personal financial needs. *Notes*: The figures stem from evaluations on a scale ranging from 1 (not at all likely) to 5 (very likely). The figure stems from Question 33 in the questionnaire.

likely". Figure 30, on the other hand, presents responses on the likelihood to use in particular such a platform to get updates on one's personal financial needs presenting the results from Question 33 that followed all the other questions that gave the participants a clearer idea of what elements the PROFIT platform could incorporate in its structure. Once again, responses were on a 1–5 Likert scale, with 1 representing "Not at all likely" and 5 representing "Very likely". As can be seen, participants are quite positive regarding using such a platform in both cases, being slightly more positive in the second case, after answering the questions regarding the platform's features and services. More experienced user groups (e.g., professionals in financial services/financial experts, collective investors/borrowers/third sector organizations, entrepreneurs), as well as parents of children scored higher regarding the likelihood to use the platform for their personal financial needs.

Table 6. Open-ended responses regarding financial education material

- Useful and easy to understand materials, surveys for improvement, useful and practical tips
- Financial basic information for children - aimed at secondary school as these skills are life skills and not the prerogative of the financial world since everybody handles money on a daily basis and we need to fin insight how to manage them. This is often a cause of poverty
- Public lectures by competent but not biased body
- Online lectures/clips (similar to courses provided by Coursera)
- Financial education seminars
- Additional web links to financial sites
- Links to financial seminars that are downloadable
- Explanatory videos
- Academic papers
- Peer-reviewed articles
- Contributions and articles of recognized experts, results of the research of the financial market, organization of workshops, seminars and courses
- A generic flow chart which may indicate best investment options after considering personal financial situations, circumstances and factors

Notes: The table stems from textual analysis of the responses to Question 34 in the questionnaire.

Table 7. Open ended responses regarding financial news/information/sentiment

- Leading magazine articles (Economist)
- Articles from local papers
- Financials News, Newspaper and Advisers
- Business section of national newspapers
- Financial news reports
- Media, information leaflets
- Exchange Rates, Commodity Prices, Stock Market Listings
- Share price updates, Index prices
- International trends, possible future outcome and past examples
- World Bank and European Bank info and forecasts, reports and rates
- Info on exchange rates, credit worthiness of investments, list of bonds in international markets
- A simpler and more usable analysis of the information published by the Malta Stock Exchange. I would like to learn how investments perform over a number of months/years
- Short and long-term provision of the stock market, raw material
- Variations in interest rates per year per country
- Deflationary trends, government spending and measure market expectations
- Time Series: currencies, commodities, country debt
- Level of trust index on locally based institutions as rated by customers/MFSA/reputable international authorities
- Indication of the critical issues identified in the simulation

Notes: The table stems from textual analysis of the responses to Question 34 in the questionnaire.

3.7   Analysis of the Open-Ended Questions

The last two questions of the questionnaire asked respondents to recommend three other resources as well as three features that they would find useful to exist in a platform that provides personal finance training.

Their responses regarding the first question were grouped in the following seven categories: Tips related to: (1) Educational material; (2) Financial news/information/sentiment; (3) Platform features; (4) Other answers; (5) Consulting-type advice - Links - Social partnerships; (6) Financial-institution/product specific; (7) Generic or other specific content. The contents which are of the highest relevance, i.e. those related to the educational material and the financial news/information/sentiment are presented in the following tables (Tables 6 and 7).

Table 8. Open-ended responses regarding suggested platform features

Response grouping	Features
Group 1: User friendliness – appeal – easy navigation	• User-friendly explanations • More user friendly • Idiot-proof • Easy access to information
Group 2: Ease of access – mobile	• Easy access • Fast platform • Quick response when interacting with the platform services
Group 3: Accessibility – localization	• Ability to print part of the content
Group 4: Privacy – security	• Security of the site platform.
Group 5: Quality of content – up to date information & notifications – Q&A forum	• Updated information • Summary of features of competitors' rated according to various benefits by users
Group 6: Communication – social network	• User network development • Networking with similar applications • Competition elements
Group 7: Financial services – Personalized support/advice	• Adverts on newspapers and TV • Financial rewards • Online loan calculator, info on advantages of passive income, importance of assets building (see Robert Kiyosaki) • Forming of various entrepreneurs clusters, forming of the various product and services offer platform, forming of buyers platform • Computational programs • Calculators for different financial products

Notes: The table stems from textual analysis of the responses to Question 33 in the questionnaire.

Finally, question 35 in the questionnaire asked respondents to recommend three other features that would make them use the platform more often. Since responses were provided in an open-ended format, they were grouped according to thematic relevance into the following seven groups and shown in Table 8:

1. User friendliness (suggestions referring to the user interface design, ease of use, friendly language, etc.) – appeal (relevant to the visual appeal, appearance, user experience, etc.) – easy navigation (ease of navigation, structure of information, etc.)
2. Ease of access – mobile (refers to fast and easy access from every device – Operating System)
3. Accessibility – localization (refers to design taking into consideration people who experience disabilities & adaptation of the platform to specific locale)
4. Privacy – security (including anonymity concerns)
5. Quality of content – up to date information & notifications – Q&A forum (suggestions relevant to the inclusion of reliable and unbiased information, regularly updated and accurate information, newsletters, daily notifications, ask questions to experts or the community, provide answers)
6. Communication – social network (features that facilitate socializing, communications and discussions with other people via fora or personal messages)
7. Financial services – personalized support/advice (having the possibility of investing, buy financial services, portfolio management, get advice and support regarding personal financial decisions)

4   Conclusions

This study presented the results from the online survey that was conducted in order to elicit user requirements and attitudes towards an online financial awareness platform, to be later used in the design of the PROFIT platform. Moreover, it provided an overview of the pertaining literature that led the authors to the delineation of the particular target user groups.

The responses in the online survey suggest that there is indeed an interest from the "bottom up" to have a trusted, reliable platform that will provide certain services towards financial awareness. In fact, what was indicative from the data analysis, is that many people do not know where to look when searching financial information on the Internet, and that there is clearly a market for online personal finance training.

On the other hand, it was found that the diversity of the sample is an important factor to take into consideration for the design of the platform; the questionnaire was distributed in six countries, all within the EU, but with different economic situations and regulation. Obviously, the platform functionalities that will be tested also have to be practically feasible to implement. Therefore, the diversity of the target audience and its motives for using and participating in such a platform are important factors that should be taken into account in the design of the platform.

Lastly, another important finding is relevant to the discrepancy between the high figures for those who do not know how to seek for information regarding changes in financial trends, and the significant but still low figure for the use of specialized websites. This finding indicates once more that the PROFIT platform is likely to address a strong existing need in the market for financial knowledge/information web tools.

Acknowledgements. This work has been supported by the EU HORIZON 2020 project PROFIT (Contract no: 687895).

References

1. PROFIT Project. http://projectprofit.eu/
2. Klapper, L., Lusardi, A., Panos, G.: Financial literacy and entrepreneurship. Global Financial Literacy Excellence Center working paper 2015-3 (2016a)
3. Brown, M., Grigsby, J., van der Klaauw, W., Wen, J., Zafar, B.: Financial education and the debt behavior of the young. Rev. Financ. Stud. **29**(9), 2490–2522 (2016)
4. Panos, G.A., Wright, R.E.: Financial literacy among Scottish students. Working paper (2016)
5. Lusardi, A., Mitchell, O.S.: The economic importance of financial literacy: theory and evidence. J. Econ. Lit. **52**(1), 5–44 (2014)
6. Lusardi, A., Michaud, P.C., Mitchell, O.S.: Optimal financial knowledge and wealth inequality. J. Polit. Econ. (2015, to appear)
7. Demirgüç-Kunt, A., Klapper, L., Panos, G.A.: Determinants of saving for old age around the world. In: Mitchell, O.S., Maurer, R., Orszag, J.M. (eds.) Retirement System Risk Management: Implications of the New Regulatory Order. Oxford University Press (2016, in press)
8. Lusardi, A., Mitchell, O.S.: Financial literacy around the world: an overview. J. Pension Econ. Finan. **10**(4), 497–508 (2011)
9. Klapper, L.F., Panos, G.A.: Financial literacy and retirement planning: the Russian case. J. Pension Econ. Finan. **10**(4), 599–618 (2011)
10. Van Rooij, M., Lusardi, A., Alessie, R.: Financial literacy, retirement planning and household wealth. Econ. J. **122**, 449–478 (2012)
11. Barber, B.M., Odean, T.: Boys will be boys: gender, overconfidence and common stock investment. Q. J. Econ. **116**(1), 261–292 (2001)
12. Panos, G.A.: 5 findings about the link between financial awareness and financial stability. EPALE blogpost (2016). http://ec.europa.eu/epale/en/blog/5-findings-about-link-between-financial-awareness-and-financial-stability
13. McPhail, K., Paisley, N., Paisley C., Panos, G.A.: Accounting in the home and school – a social constructivist perspective. University of Glasgow, Working paper (2016)
14. Nolte, S., Schneider, J.C.: Don't Lapse into Temptation: A Behavioral Explanation for Policy Surrender. Working paper (2015)
15. Lusardi, A., Tufano, P.: Debt Literacy, Financial Experiences, and Over indebtedness. National Bureau of Economic Research Working Paper 14808 (2009)
16. Gathergood, J.: Self-control, financial literacy and consumer over-indebtedness. J. Econ. Psychol. **33**(3), 590–602 (2011)
17. Van Rooij, M., Lusardi, A., Alessie, R.: Financial literacy and stock market participation. J. Financ. Econ. **101**(2), 449–472 (2011)

18. Klapper, L.F., Lusardi, A., Panos, G.A.: Financial literacy and its consequences: evidence from Russia during the financial crisis. J. Bank. Finance **37**(10), 3904–3923 (2013)
19. Panos, G.A., Wright, R.E.: Financial literacy and attitudes to the Scottish referendum. Working paper (2016)
20. Panos, G.A., Wright, R.E.: Financial literacy and attitudes to the EU and referendum. Working paper (2016)
21. Yamaguchi, S.: Tasks and heterogeneous human capital. J. Labor Econ. **30**(1), 1–53 (2012)
22. Gathmann, G., Schönberg, U.: How general is human capital? a task-based approach. J. Labor Econ. **28**(1), 1–49 (2010)
23. Autor, D.H., Handel, M.J.: Putting tasks to the test: human capital, job tasks and wages. J. Labor Econ. **31**(2), S59–S96 (2013)
24. Montagnoli, A., Moro, M., Panos G.A., Wright, R.E.: Financial literacy and attitudes to redistribution. Working paper (2016)
25. Moore, D.: Survey of Financial Literacy in Washington State: Knowledge, Behavior, Attitudes, and Experiences. Washington State University Social and Economic Sciences Research Center Technical report 03-39 (2003)
26. Lusardi, A.: Americans' Financial Capability. National Bureau of Economic Research Working Paper 17103 (2011)
27. Disney, R., Gathergood, J.: Financial literacy and consumer credit portfolios. J. Bank. Finance **37**(7), 2246–2254 (2013)
28. Clark, R.L., Matsukura, R., Ogawa, N.: Low fertility, human capital, and economic growth: the importance of financial education and job retraining. Demographic Res. **29**, 865–884 (2013)
29. Lusardi, A., Schneider, D., Tufano, P.: Financially fragile households: evidence and implications. In: Brookings Papers on Economic Activity, pp. 83–134 (2011)
30. Bodnaruk, A., Simonov, A.: Do financial experts make better investment decisions? J. Finan. Intermediation **24**(4), 514–536 (2015)
31. Demirguc-Kunt, A., Klapper, L., Singer, D., van Oudheusden, P.: The global findex database 2014: measuring financial inclusion around the world. Policy Research working paper; no. WPS 7255, World Bank Group, Washington, D.C (2015)
32. World Bank: Global Survey on Consumer Protection and Financial Literacy: Oversight Framework and Practices in 114 Economies. The World Bank, Washington, DC (2014)
33. Freeman, R.B.: The exit-voice tradeoff in the labor market: unionism, job tenure, quits, and separations. Q. J. Econ. **94**(4), 643–673 (1980)
34. Panos, G.A., Theodossiou, I.: Reciprocal loyalty and union mediation. Ind. Relat. J. Econ. Soc. **52**(3), 645–676 (2013)
35. Hvide, H.K., Panos, G.A.: Risk tolerance and entrepreneurship. J. Financ. Econ. **111**(1), 200–223 (2014)
36. Cohn, A., Engelmann, J., Fehr, E., Maréchal, M.A.: Evidence for countercyclical risk aversion: an experiment with financial professionals. Am. Econ. Rev. **105**(2), 860–885 (2015)
37. Klapper L., Lusardi A., Panos, G.A.: Financial literacy and trust in financial institutions. Working paper (2016)
38. Montagnoli A., Moro M., Panos G.A., Wright, R.E.: Financial literacy and political attitudes. Working paper (2016)
39. Von Gaudecker, H.M.: How does household portfolio diversification vary with financial literacy and financial advice. J. Finan. **70**(2), 489–507 (2015)
40. Klapper L., Lusardi A., Van Oudheusden, P.: Financial literacy around the world: Insights from the Standard & Poor's ratings services global financial literacy survey. S&P Report (2015)

41. Islam, R. (ed.): Information and Public Choice: From Media Markets to Policy Making. The World Bank, Washington, DC (2008)
42. Calcagno, R., Monticone, C.: Financial literacy and the demand for financial advice. J. Bank. Finance **50**, 363–380 (2015)
43. Panos, G.A., Wright, R.E.: Financial literacy and attitudes to immigration. Working paper (2016c)

A Reputation-Based Incentive Mechanism for a Crowdsourcing Platform for Financial Awareness

Aikaterini Katmada, Anna Satsiou$^{(\boxtimes)}$, and Ioannis Kompatsiaris

CERTH-ITI, Thessaloniki, Greece
{akatmada, satsiou, ikom}@iti.gr

Abstract. This paper presents the design of an incentive mechanism for the so-called PROFIT platform, a crowdsourcing (CS) platform that seeks to promote financial awareness and capability. More specifically, a reputation-based incentive scheme with gamification and social elements, which offers a mix of both implicit and explicit rewards to the most contributive users of the platform, is being proposed here. The incentive mechanism has been designed in a way to appeal to the various different motives of the target users of the platform, in order to encourage their active participation, sustain their interest and engagement, and promote good quality contributions. After reviewing the relevant work regarding incentive mechanisms in CS platforms, we present the rationale behind the design of the proposed scheme, following a five-step approach and presenting the novelties that we introduce, and, lastly, we conclude on some final remarks.

Keywords: Crowdsourcing · Incentives · Financial awareness · Reputation systems · Gamification · Collective Awareness Platforms for Sustainability and Social Innovation (CAPS)

1 Introduction

Recent advances in the field of communication technologies have transformed the way we think and live, and have also led to a major increase in the human knowledge capital [1]. Indeed, contemporary times are commonly referred to as the "information age", mainly characterized by the *"diffusion of information, as well as Information and Communications Technologies"* [2]. Connectivity, sharing of knowledge and collaboration among users are now facilitated by Web 2.0 (the so-called "Participative Web"), and applications such as social networking sites, peer-to-peer (P2P) media sharing utilities, blogs, and wikis [2] that emphasize on users being active contributors of content, and not just viewers and consumers of information [1]. Moreover, it is argued that these applications cater to the needs of today's learners, by giving them the opportunity to collaborate, network, and customize them [3]. In that context, users of various educational levels, e.g. learners, educators, and experts, are able to collaborate and take active part in the knowledge creation process. This process is also supported by the so called crowdsourcing (CS), a practice that emerged with the development of Web

© Springer International Publishing AG 2016
A. Satsiou et al. (Eds.): IFIN and ISEM 2016, LNCS 10078, pp. 57–80, 2016.
DOI: 10.1007/978-3-319-50237-3_2

2.0. Crowdsourcing is known as a distributed online process that harnesses the participation of the crowd for the achievement of particular tasks [4], and has been successfully used for educational, decision making, and knowledge creation purposes [5].

The EU-funded "PROFIT" project [6] aims at developing a CS platform that would promote financial awareness and capability. Financial awareness has been deemed essential to the society, since it can support citizens in taking informed financial decisions which will lead to more responsible and prudent personal financial behavior and, ultimately, to more active forms of participation and citizenship. Although there is nowadays a large amount of financial information, easily available due to the aforementioned technological developments, the documented human cognitive limitations of processing large volumes of information, in conjunction with the widespread financial illiteracy, indicate that there is a need for special financial awareness tools. Towards this goal, the PROFIT platform will incorporate the following functionalities/tools: (a) specialized financial education toolkits available to the wider public; (b) advanced crowdsourcing tools that process financial data, and also extract and present collective knowledge; (c) advanced forecasting models exploiting the market sentiment to identify market trends and threats; (d) novel personalized recommendation systems to support financial decisions according to the user's profile (financial literacy level, interests, demographic characteristics); and (e) incentive mechanisms to encourage the active involvement of citizens through many different channels, such as posting of financial articles, participating in Q&A forum, voting in relevant polls, etc.

This paper deals with the last one of the aforementioned functionalities, i.e., incentive mechanisms to attract users' attention and sustain their engagement with the platform. In order to achieve PROFIT's goal and promote financial awareness, users' participation and contributions are critical to keep the platform alive and load it with articles, questions, answers, and polls. Other users can see, assess, rate, comment, react to this content and provide relevant feedback, leading to a fruitful creation and exchange of new knowledge in the financial domain. Therefore, in order to achieve this, we propose a novel reputation-based incentive scheme with integrated gamification elements, combining both implicit (e.g. social status) and explicit rewards (e.g. tangible rewards) in a way that could appeal to the different target users of the PROFIT platform. In the following section (Sect. 2) we delve into related work in the area of incentive mechanisms for CS platforms, explaining the novelties that we introduce. In Sect. 3, we present our incentive mechanism through a five stage design approach in order to meet the particular needs of the PROFIT platform, and in Sect. 4 we conclude the paper with some final remarks.

2   Related Work

2.1   Theoretical Background - User Motives and Incentives in CS Platforms

The success of CS platforms depends greatly on the continuous participation of the users and their sustained engagement, and that explains the amount of studies focusing on user motives for participating in CS platforms (e.g. [7, 8]). Some common user motives that were identified include altruism, learning/personal achievement, social motives, self-marketing, direct compensation, and enjoyment [5, 7, 9, 10]. They can be activated either by intrinsic incentives, e.g. social status and respect by others [7], or by extrinsic incentives, such as payment [5]. In our previous study [11] we investigated the connection between different user motives and corresponding incentives, as well as appropriate incentive mechanisms that could be designed in order to trigger the particular incentives. The identified incentive mechanisms were organized in four main categories, consisting of the following: (a) reputation systems, (b) gamification, (c) social incentive mechanisms, and (d) financial & career rewards. A reputation system is used to identify the most contributive (reputed) users through the implementation of a reputation metric. It usually involves some implicit or explicit rewards for users with high reputation, as well as penalties for users with very low reputation. These rewards may span from badges and social status, to career opportunities and financial awards; therefore, reputation systems are usually combined with one or more of the rest of the incentive mechanisms mentioned above. In PROFIT, we made a careful combination of all four aforementioned incentive mechanisms, in order to appeal to the different motives of the various user communities, and sustain their interest and engagement in the platform. To the best of our knowledge, no other CS platform has combined all of the above mechanisms.

2.2 Incentive Mechanisms in CAPS Platforms

PROFIT is part of the initiative "Collective Awareness Platforms for Sustainability and Social Innovation" (CAPS), which is focused on the design of online platforms that will *create awareness of sustainability problems and offer collaborative solutions based on networks* [12]. At the moment, a total of 12 projects from the first call are already running since 2013, while 22 new projects, including PROFIT, have commenced their activities this year. An open problem of CAPS projects is how to reach to their user communities; therefore, the design and incorporation of appropriate incentive mechanisms play a significant role. It is of interest, though, to overview what has been implemented in this context, so far.

The incentive mechanisms employed until now consist mainly of social incentive mechanisms, appropriate feedback (e.g. activity history, advanced visualizations) that support and motivate users who participate for altruistic or self-accomplishment reasons (e.g. learning), and, less often, simplistic reputation systems. Gamification elements or direct monetary/career rewards were not identified. An example would be DebateHub [13], a tool that facilitates the sharing and discussion of ideas between users, developed by the CAPS project Catalyst [14] which aims at improving collaborative knowledge creation. In order to achieve prioritizing of contributions and eventually promote the best ideas, a simple rating mechanism was incorporated: users have the option to upvote and downvote ideas. Social incentives mechanisms (e.g. user discussion groups, follow other users), as well as a visualization dashboard consisting of summary analytics and attention mediation feedback were implemented; however, there is no reputation system implemented that could assess users' credibility and behavior on the platform.

Similarly, the Wikirate CAPS project [15] has created a web platform for collaborative creation and sharing of knowledge on company behavior. Users can contribute information and have access to data and visualizations reflecting company behavior in order to compare and rate these companies. There are some social incentives mechanisms (e.g. profile pages, follow other users), and users have the option to upvote or downvote other users' posts. However, there is no reputation system implemented here either. There are plans to incorporate several gamification elements, e.g. badges, in the near future [16].

A simple reputation system in which users accumulate points according to the upvotes their posts receive from other users was implemented in the EnergyUse platform [17]. The particular platform allows citizens to discuss about energy saving with users of electricity monitors aiming at engaging them in fruitful energy debates and promoting positive behavioral change towards reduced energy consumption. It was created by the CAPS project DecarboNet [18], which is investigating the potential of social platforms in mitigating climate change. Apart from social incentives mechanisms (comment on posts, discuss with other users), EnergyUse also incorporates a basic reputation system: users can upvote posts and answers, and accumulate reputation points accordingly. It is mentioned on the platform that awards for the users will be incorporated soon.

By the time the particular paper was written, the authors did not encounter any other CAPS projects that had incorporated reputation-based incentive mechanisms or gamification elements into their platforms. However, some of them included social incentives in the form of offline meetings and social events in order to inform the wider public and increase users' interest and engagement on the platform. For example, CAP4Access [19], a CAPS project which aims at developing methods and tools for the collective gathering and sharing of spatial information for improving accessibility, has also created a web "communication platform", which features activities and meetings initiated by the CAP4Access partners, as well as the work of other activists and organizations working for accessibility in Europe.

2.3 Simple Reputation Schemes Used in Commercial CS Platforms Versus Sophisticated Ones Proposed in the Literature, and the PROFIT's Approach

As far as incentive mechanisms in other CS platforms (commercial and research projects) are concerned, these are usually a combination of reputation systems with other incentive mechanisms, i.e. gamification and financial rewards, as already mentioned. The reputation metric can be based upon various different methods, spanning from simple summation of ratings to fuzzy logic and probabilistic models [20]. Simple reputation systems are preferred in commercial CS platforms, since they are easily applicable and they are not that computationally expensive. Some indicative examples of such reputation systems in CS platforms, combined with social incentive mechanisms, include those implemented in CS news websites, such as Reddit, Slashdot, and Hacker News. The reputation metric here is mostly based on accumulation of user points ("Karma"), according to the upvotes and downvotes their contributions received by other users. Simple reputation systems are also implemented in auction and e-commerce websites, such as Amazon and eBay, where buyers and sellers can rate each other on a scale 1–5 and leave comments. The reputation of each user is directly linked to career and financial rewards and can affect them favorably or negatively. Similarly, In online marketplaces, such as Amazon Mechanical Turk, a user's reputation score usually reflects her trustworthiness and her abilities and skills, since it is often calculated as the percentage of her successfully completed tasks.

On the other hand, more advanced and fair reputation schemes have been proposed in the pertinent bibliography. The main concern here is how to increase the system's robustness against user manipulation. As a result, complexity as well as computation burden may be here increased contrary to simpler reputation schemes. For example, Whitby et al. [21], proposed a Bayesian reputation system based on [22]. It filters unfair ratings by excluding or giving low weight to presumed unfair ratings, based on the assumption that they can be recognized by their statistical properties [21]. Similarly, Dellarocas in [23] focuses on a set of mechanisms based on controlled anonymity and cluster filtering that can be integrated into a reputation system to address unfair user behavior (e.g. unfairly high or low ratings).

At PROFIT, we followed a middle solution by designing a reputation scheme that is neither overly simplistic, nor too complicated in order to be easily implemented in a practical platform, incorporating a reputation mechanism that tracks not only contributive users but also those particular users that provide high quality contributions, and rewards them accordingly. It differentiates from other reputation systems we overviewed, by using two distinct reputation metrics, as well as a time window, in order to create a more robust solution for incentivizing users, as we will elaborate in the next sections. Additionally, a unique combination of the aforementioned incentive mechanisms are going to be incorporated, in order to create an integrated incentive system that would appeal to different user groups.

3 Design of the System

Gamification in the PROFIT platform will act as an additional service layer of the reputation system presenting the user with game design elements, such as points, badges, time constraints, levels, leaderboards, [24]. Therefore, several gamification conceptual frameworks were identified and studied. Some prominent examples include Di Tommaso's gamification framework [25], Chou's "Octalysis" framework [26], Brito's et al. "GAME" framework [27], and Werbach's six-step gamification framework [28]. Based on a combination of the above, the authors concluded on five steps that would be most suitable to follow, in order to gamify the particular platform: (a) delineation of the platform objectives and desired user's behaviors; (b) description of target users and investigation of their motives to participate in the PROFIT platform; (c) identification and design of appropriate incentive mechanisms that appeal to the targeted users' motives and trigger the desired behaviors; (d) inclusion of measures to sustain user engagement; and (e) definition of the evaluation strategy. These steps are going to be presented in the following subsections.

3.1 Objectives of the Platform and Target User Behaviors

The PROFIT platform aims at promoting the financial awareness of EU citizens and other financial market participants. Towards this goal, it is intended to trigger users' interest on the financial domain, help them improve their financial literacy levels, and motivate them to participate in relevant discussions and interactive activities in the platform, so as to raise awareness on different financial issues. Such activities include: self-assessment of their own financial knowledge levels through the financial literacy tests of the platform, upload of financial-related articles, posting of related questions or answering questions, participating in polls or creating their own, rating or commenting on other users' posts or polls, rating or commenting on the platforms' content, provide external or their own educational material, edit other users' provided educational material, etc. Additionally, what is needed is to not only encourage users' participation and contributions on the platform, but also to motivate good quality contributions that could support the platform's goal, as well as to sustain this desired behavior over time.

To reach the aforementioned goals, it is of utmost importance to communicate in the right way what the platform wants to achieve, and promote a sense of ownership of this goal among the platform participants so that they can understand and deeply share the goal. This can be supported by providing users with appropriate and timely feedback regarding the impact of their contributions on reaching this goal (e.g. through interactive visualizations, graphs, etc.), and by promoting collaboration and social interactions among users. Special feedback should also be provided to newcomers, by giving them simple guidelines on how they can participate and contribute, highlight what they can achieve from their participation, and provide them with certain rewards for their initial contributions.

3.2 Description of Target Users and Investigation of Their Motives

The proposed platform will address all kinds of users, from people with little/no financial knowledge to more financial literate users, regardless of age, gender, financial background, etc. The authors, based on relevant research [33] and the feedback from the PROFIT User Forum committee consisted of financial institutions, entrepreneurs, government bodies, educational institutions, banks' customers and other potential customers and/or final users of the PROFIT platform, concluded on the following target user groups of the proposed platform, based on the demographic category that would better reflect their financial status and needs:

- UG1: Entrepreneurs/latent entrepreneurs/social entrepreneurs/self-employed,
- UG2: Elderly/retirees/pre-retirees,
- UG3: Migrants/Members of an ethnic minority,
- UG4: Children/Parents of young children,
- UG5: Customers: Indebted/Over indebted households,
- UG6: Customers: Investors/Potential investors/Depositors,
- UG7: Unemployed/trainees,
- UG8: Active citizens/taxpayers,
- UG9: Mortgage owners/home owners/first-time buyers,

- UG10: Professionals in financial services/financial experts,
- UG11: Government executives and political-party members/local authorities,
- UG12: Collective investors/borrowers/third sector organisations.

In order to investigate the target users' attitudes towards a financial awareness platform and understand their motives for participating and contributing to such a platform, both literature survey and a large scale online survey was conducted. The online questionnaire[1] comprised 35 questions related to the demographic information of the participants, their financial knowledge and status, their attitude towards using technology and Internet, their ICT skills, and their motives for participating and contributing to such a platform. It was available for completion online in six languages (English, French, Italian, Greek, Croatian, and Slovenian). The English version questionnaire is available at http://projectprofit.eu/material/#tab-id-1.

The questionnaire was completed by 494 people of varying age, nationality, and education/professional background. 50.2% of the completed questionnaires were in English, 24.3% were in Greek, 7.1% were in Croatian, 6.7% were in Slovenian, 6.1% in Italian, and 5.7% in French (Fig. 1). The majority (296 or 59.9%) participants were male and 198 (40.1%) were female (Fig. 1). The youngest person that participated in our survey was 18 years old and the oldest was 85 years old. Overall, 40.7% of the participants were between 28 and 43 years old, 30.1% were 44–60 years old, 16.7%

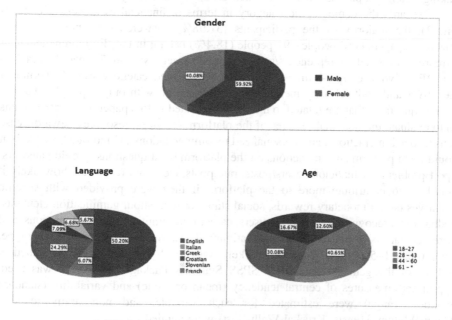

Fig. 1. Gender, language and age of respondents

---

[1] The realization of the PROFIT's online questionnaire has received the ethical approval from the Bio-Ethics Committee of the Centre for Research and Technology Hellas with REF No: ETH. COM-19.

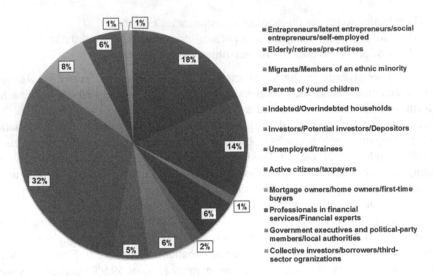

Fig. 2. Demographic category of respondents

were older than 61 years old, and only 12.6% of the participants were 18–27 years old, making young people the least represented group in our survey (Fig. 1).

As regards the demographic category in terms of financial awareness and needs (Fig. 2), the majority of the participants (31.98%) answered that they are active citizens/taxpayers (158 people). 91 people (18.4%) belong in the 'Entrepreneurs/latent entrepreneurs/social entrepreneurs/self-employed' category, and 71 people (14.4%) were 'Elderly/retirees/pre-retirees'. The least represented category was 'Government executives and political-party members/local authorities' with only 3 people (0.6%).

The questions that are related to the work presented in this paper asked participants (a) to evaluate the potential features of the platform (services, resources, rewards, ease of use, social interactions, and personalized recommendations), (b) to rate how likely it would be to perform specific actions on the platform (post questions, provide answers, report problems or malicious users/posts, rate posts, etc.), and (c) to rate how likely it would be to contribute more to the platform if they were provided with specific incentives (small monetary rewards, social status & reputation, gamification elements, feedback and recognition from other users, career opportunities, social interactions, and more accurate personalized recommendations). These questions were of Likert type, using a scale 1–5 where 1 = "Not at all likely" and 5 = "Very likely". For the statistical analysis of the gathered data, IBM SPSS Statistics Package version 20 was used. Appropriate measures of central tendency (mean or mode) and variability (standard deviation, range) were estimated for each variable, and non-parametric tests (Mann-Whitney U tests, Kruskal-Wallis test) were applied.

Respondents were mostly positive (mode 4) regarding the proposed features of the platform. The feature with the most positive answers was Ease of use: 69.8% of the participants answered "Very likely" and "Quite likely". Respondents were positive (mode 4) regarding participating in polls and providing feedback to the platform, and neutral (mode 3) regarding posting questions, providing answers, and rating posts.

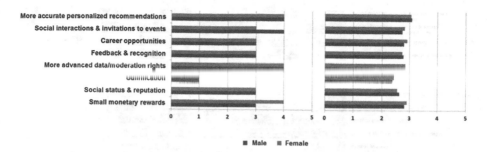

Fig. 3. Incentives according to gender (mode-left, mean-right)

Lastly, incentives that seem to be the more appealing according to the results are *social interactions & invitations to social events* according to one's contributions, *access to more advanced data & moderation rights*, and *more accurate personalized recommendations* (mode 4).

We also analysed the participants' responses according to their gender, age, and user group. What we saw is that incentives do not present significant differences between males and females that both seem to be more interested in receiving *more accurate personalized recommendations*, with females showing a slightly higher than males interest in *career opportunities* and *small monetary awards*, while males showing a slightly higher interest than females in gaining *access to more advanced/moderation data*, as well as *social status, reputation* and *recognition* from the community (Fig. 3). Nevertheless, both males and females seem equally interested in using the PROFIT platform, as 179 (60.9%) of male participants and 103 (52.8%) of female participants stated that it would be quite/very likely that they use the platform.

Regarding differences between different ages target groups, we saw that in general younger people (age 18–43) were more interested in the different incentives, with a strong preference on *career opportunities, small monetary rewards, access to more accurate personalized recommendations* and *social interactions*, while their older counterparts (age 44+) show some interest only for the *more accurate personalized recommendations* and access to *more advanced data/moderation rights*. The majority (mode) of the youngest group (age 18–27) was also positive regarding the *gamification elements* and the "*social interactions and invitation to events*"; the majority of the 28–43 age group was also positive on receipt of *feedback and recognition* according to

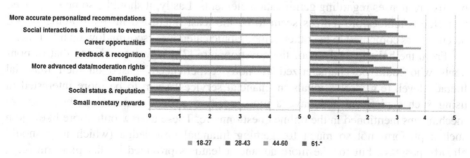

Fig. 4. Incentives based on respondents' age (mode-left, mean-right)

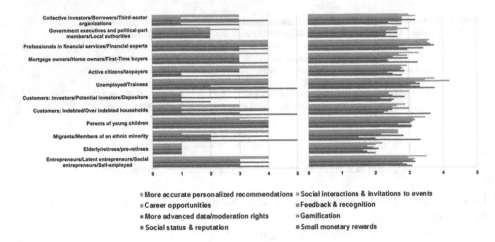

Fig. 5. Incentives based on demographic category (mode-left, mean-right)

their contributions, the majority of the age 44–60 group was positive regarding the *more accurate personalized recommendations* and access to *more advanced data/moderation rights,* while the majority of the age 61+ group was negative in terms of all listed incentives (Fig. 4).

Lastly, as depicted in Fig. 5, there were also identified differences which stem from the respondents' financial status and needs, as these are reflected on their chosen demographic category. *Small monetary rewards* seem to be the strongest incentive for the majority of unemployed/trainees, indebted/over indebted households and migrants/members of ethnic minorities. *Career opportunities* also constitute a strong incentive for unemployed/trainees, while *social interactions,* gaining *social status and reputation* as well as *feedback and recognition* from the community also are of interest to them. Interestingly, these same incentives are also a motivating factor for professionals in financial services/financial experts, who seem to be interested in self-marketing and socializing, being quite positive towards *feedback* and *recognition according to their contributions, social status and reputation, social interactions* on the platform, and *invitations to social events.* Parents of young children, who may be concerned regarding the financial education of their children and enjoyable ways to motivate them in being informed in this area, are among the groups who gave the most positive responses regarding gamification elements. Lastly, it should also be mentioned that elderly/retirees/pre-retirees seem to be the least interested in the aforementioned incentives as compared to the other demographic categories, scoring low on all of them.

From the holistic analysis of the questionnaire [33] it was concluded that respondents who could be characterized as more "experienced" based on their financial literacy level, (e.g. professionals in financial services) seem to be more interested in using such a platform and have a more positive attitude concerning the incentive mechanisms mentioned in the online questionnaire. These users would more likely join such a platform not so much for gaining financial knowledge (which they should already possess), but for the more advanced features provided by the platform (e.g., financial forecasting based on market sentiment) that could be of interest to them. Moreover, these users could also be motivated by having the opportunity to contribute their knowledge either for altruistic purposes (i.e., contributing financial information for the good cause of promoting financial awareness globally) or for self-marketing and gaining social status/reputation by promoting their skills through their activities on the platform. Therefore, their dominant motives for participating comprise altruism, self-marketing, social motives, and direct compensation.

"Moderate" users in terms of financial literacy (e.g. insurance/mortgage/home owners and first-time buyers) are potentially motivated by their will to learn more about specific financial matters and get relevant recommendations, as well as socialize and network with other users of the platform. Lastly, users that may be characterized as less financial literate or less financially independent (e.g. unemployed/trainees) may participate in order to get informed, ask questions, socialize, explore career opportunities, or gain small monetary rewards (e.g. prizes).

As a concluding remark regarding user incentives, we argue that the motives and reasons for participating in the PROFIT platform vary according to the age, and user group of the participants. We can also expect that their participation levels and patterns on the platform would differ greatly. Therefore, the incentive mechanism that will be incorporated into the platform should make a careful combination of different incentive mechanisms in order to appeal to the majority of them and be effective for all target users.

Finally, is should also be mentioned that participants in the survey were also given the opportunity to suggest features for the PROFIT platform. Most of their answers were focused on the importance of an appealing user interface and an easy to use web platform/application. Nonetheless, among their suggestions were also Question & Answer Forums, private and instant messages, awards, opportunities to socialize with other users and ways to exchange knowledge with them.

3.3   How to Trigger User Motives - Incentive Mechanisms Applied and Designed

Based on a previous work [11], as well as the conclusions from the results of the online survey mentioned above, the authors decided that in order to correspond to the motives of all the aforementioned user groups, the PROFIT platform should combine a variety of different incentive mechanisms (reputation system, gamification, social incentive mechanisms, and financial rewards). More specifically, a reputation system would correspond to user's self-marketing motives by giving them the opportunity to showcase their social status and demonstrate their skills. Users' altruism can be triggered by the opportunity to contribute for a good cause (in that case promotion of financial awareness to other users), as well as feedback (through appropriate design elements like visualizations, graphs, etc.) regarding the impact of personal contributions to the platform community. Gamification mostly appealing to the younger age groups elicits user motives such as enjoyment, intellectual curiosity, learning and personal achievement, by allowing the most contributive users to unlock new features (e.g., access to more advanced functionalities, moderation rights on the platform, etc.), gain rewards (e.g. badges representing the user's achievements), and upgrade their level (e.g. avatar changing accordingly to represent user's contribution levels). However, elements that may raise unnecessary competition (e.g., leaderboards) are going to be avoided in the particular platform, or used under certain conditions. Social incentives mechanisms, such as "follow" and "conversation" functionalities and social feedback (comments, ratings, etc.), appeal to the users' social motives by allowing them to present a good social image according to the values of the online community and gain appreciation by others, while they can also encourage altruistic actions. Lastly, suitable incentives for direct compensation, such as discounts on products and prizes, will also be included in the platform. In what follows, the incentive mechanisms applied are being presented in more detail.

Reputation System: A Way to Measure User Progress. In order to measure user progress, two distinct reputation metrics are going to be implemented; one to reflect the participation level of the user (R_p), and one to reflect the quality of the provided contributions (R_q). The reputation metrics will provide feedback to the users as to what extent they are engaged with the intended behaviors. These two reputation metrics would be visible on the user dashboard, alongside with visualizations of user activity. Having two separate reputation metrics allows to distinct contributive users from those users who are not only frequent contributors, but they also provide contributions of a high quality. It is important to track both the number of contributions as well as the level of their quality, so as to particularly promote very active users with high quality contributions. In the following, we explain how the two reputation metrics are calculated.

More specifically, users will be able to gain points for each of their actions on the platform, according to their importance for the platform purposes. For example, a user will be awarded 15 points each time she posts financial information on the platform, each time she adds new educational material, and each time she provides an answer to a financial related question posted by another user (Table 1). The participation level of the user (R_p) is then judged by the sum of her accumulated points on the platform. Actions that promote user collaboration and are beneficial to the community are awarded with more points (e.g. answer to other users' questions; participate in polls created by other users, etc.). Moreover, in order to promote high quality contributions, we encourage users to rate other users' posts on a scale 1 to 5, by rewarding them with a high number of points. Lastly, users are encouraged to assess and (re) assess their literacy level by taking the related test (15 points); in this way, the platform can provide them with more suitable information for their level and assess its success by tracking the users' financial literacy improvements.

On the other hand, the quality-based reputation metric is going to be recalculated on a weekly basis, in order to reflect the latest behavior of the user, based on the following formula:

$$R_q^{u_x} = \sum_{\forall p_{u_x}, u_i} r_{p_{u_x}}^{u_i} \cdot R_q^{u_i} / \sum_{\forall p_{u_x}, u_i} R_q^{u_i} \tag{1}$$

Table 1. Points awarded to users according to their actions

Actions awarding points	Points
Invite a friend through social media	+2
Write and post financial information	+15
Upload financially-related content (created by others)	+10
Post a question	+5
Answer to a question posted by another user	+15
Create a poll	+5
Participate in a poll	+8
Post a comment on another user's contribution	+5
Post a comment on educational material provided by the platform	+7
Share own post on social media	+5
Share another user's post on social media	+7
Rate on a scale 1–5	+8
Flag inappropriate post	+5
Share another user's post on social media	+7
Add new educational material	+15
Edit other user's educational material	+10
Take financial literacy test	+15

Where, $R_q^{u_x}$ is the quality-based reputation of the user u_x, p_{u_x} is a particular post of the user u_x, and $r_{p_{u_x}}^{u_i}$ is the rating provided by user u_i for the post p_{u_x} and can take discrete values between 1 and 5. Thus, the quality-based reputation of a user is calculated based on the weighted average of the ratings her posts (over the week) received by other users, where the weights are the quality-based reputations of these other users. In this way, ratings by more reputed users weight more to the calculation of the respective user quality reputation. R_q can take values within [1, 5]. Newcomers' default R_q is set to 3 and can be adapted each week according to the user's contributions and ratings they received. In case a user has not contributed anything in the platform within a week, R_q remains the same as in previous week, while the lack of contributions within this week is reflected in the participation-based reputation R_p. By re-calculating the quality reputation metric of users each week, we protect the system against potential malicious or misbehaving users that seek to gain reputation fast and then decrease their participation levels or the quality of their contributions and ratings [29] with no costs at the benefits they can receive from the platform; any behavior change will be promptly tracked by the system and will be reflected on the user accumulated points (R_p) and/or quality reputation (R_q) scores, respectively, with the relative consequences on the users' rewards and rights, as will be explained in the gamification elements paragraph below. These two reputation metrics provide users with information regarding their progress and will be visible on the user profile page (Fig. 8).

The proposed reputation system also provides certain awards or penalties for users according to their reputation scores. Awards vary from social status gained in the platform, depicted with the use of certain gamification elements, to moderation rights in

Temporarily banned   Permanently banned

Fig. 6. User avatar for banned users

the platform, as well as tangible rewards (e.g. small monetary prizes), as described below. Penalties affect the users' social status negatively. They consist of changing the avatar that appears below the "Badges" and indicates a user's level to a specific icon, "warning" other users that this user's posts were flagged as inappropriate by moderators (Fig. 6); giving administrators the right to ban malicious/misbehaving users so that they cannot post and comment on the platform for a certain period of time (e.g. a week); and permanently expelling users with a history of 6 or more repeated bans from the platform.

Gamification Elements. Gamification elements that are going to be included in the design of the PROFIT platform comprise user avatars, levels, special achievements, progress charts and badges. More specifically, by increasing their participation based reputation (R_p), users unlock levels and gain more rights on the platform. There are six levels on the platform: newcomer, experienced, casual, master, expert, moderator, as well as six corresponding icons (Fig. 7). The icon corresponding to a specific user's level is visible on a user's profile below her acquired badges, as mentioned before. Users begin at the newcomer level with basic functionalities and avatar, and they move up to the casual level, where they can create their own community polls, when their R_p reaches 300 points. After that, they may unlock the experienced level when their R_p reaches 1000 points, the master level when R_p reaches 2000 points, etc. (Table 2). It should be noted that users of top level which have been given moderation rights on the platform are going to be able to hide inappropriate content. In all above cases, users can upgrade their level, only if their quality reputation scores (R_q) are above 3. Thus, a user

New   Casual   Experienced   Expert   Master   Moderator

Fig. 7. User avatars according to level

Table 2. Levels & privileges according to level

Level	Name	Minimum points (R_p)	Privileges/functionalities unlocked
1	New	1	• Basic avatar
			• Search for articles and add to watch list
			• Follow other people
			• Post question
			• Answer questions
			• Share to social media
			• Participate in polls
			• Rate other users' contributions
			• Rate platform's material
2	Casual	300	• Avatar changes
			• Comment on other users' contributions
			• Comment on platform's material
			• Create poll
3	Experienced	1000	• Avatar changes
			• Get more personalized recommendations on users/articles
			• View position on private leaderboard
			• Participate in competitions for FEBEA tangible rewards
4	Expert	2000	• Avatar changes
			• Provide additional educational material
			• Flag inappropriate posts to administrators
5	Master	4000	• Avatar changes
			• Help translate the site in other languages
			• Edit educational material and/or others' posts
6	Moderator	8000	• Avatar changes
			• Access to moderators and initiators (PROFIT partners) forum
			• Moderation rights:
			– Hide/remove inappropriate content
			– More statistics on accounts, traffic patterns, preferences, visitors
			• Organize local user meetings
			• Propose and coordinate the new PROFIT directions

may advance to a particular level depending both on the amount of her contributions (R_p) as well as the quality of them (R_q).

Nevertheless, users may also lose their status and move down levels, if their weekly quality reputation is bad or they have remained inactive for a prolonged period of time. By implementing this time window, we are trying to sustain their interest and activity on the platform. Moreover, we are also trying to prevent and deter misbehaving or malicious users as explained above. More specifically, for experienced level users and above, if their quality reputation scores (R_q) falls under 3, or their accumulated points for a week are zero, 500 points will be subtracted from their participation reputation (R_p).

Additionally, users will also be presented with illustrated avatars that change according to the user level, as well as badges to indicate that they have unlocked various special achievements (Table 3). Some special achievements include the "Best-author award": given to the author of the weekly featured article (article with highest rating), the "Loyalty award": awarded to users that make at least one contribution per day for a month, the "Exceptional Contribution Award", awarded to users whose articles/answers rated over 3/5 at a month's period, etc. These special achievements are acting as milestones and their role is to reward active users with exceptional contributions on the platform. There are also badges to encourage newcomer's participation, like the "First article award" given to the users that posted their first article, the "Sociable award" given to the users who completed the "about me" profile section, and uploaded a profile photo, and the "Eager-to-learn award": given to the user who posted the most questions (per month). Last, there are badges given to reward other kind of desired behavior in the platform, like reporting of platform bugs (Bug-hunter award), and improvement of financial literacy level (Financial literacy improvement award). For all achievements listed in Table 3, users will not only be awarded with badges that are permanently showcased on their profiles as small icons (Fig. 8), increasing their social status, but also with some extra "bonus" points. Some badges and the respective points can be gained more than once; in this case a number placed next to the respective badge indicates how many times this badge has been awarded to a certain user.

Social Incentive Mechanisms. It has been shown [30] that people learn by social observation and tend to act like people they observe even without external incentives. In particular newcomers can become more aware of the platform functionalities and how these can be used, when observing the contributions of their friends. On the other hand, the feedback received on one's contributions can act as a motivating factor for even more contributions [30]. Therefore, various such kinds of social incentive mechanisms were incorporated in the platform, including elements that facilitate social interactions, such as the functionality to follow other users or to comment on others' posts, to exchange private messages with other users, and to share posts on social media. Public leaderboards that show the ranking of users according to their scores are going to be excluded from the design of the platform, in order to avoid competitive behaviors that may deter newcomers and low ranked users. Instead, users of level 3 (experienced) and above are going to be presented with private leaderboards on their (private) activity dashboards (Fig. 9). Using their private activity dashboard, users will be able to access, among others, lists with their followers and the users they follow, as well as any messages they have exchanged. Users will also receive recommendations on other similar users that they could follow, according to their interests. Lastly, they will be prompted to invite their friends to the platform using their social media accounts by rewarding them with 2 points for each invitation.

Table 3. Examples of special awards

Examples of special awards (achievements)	Bonus points & reward
First article award: after creating an article for the first time	• Special icon on user profile page • Achievement showcased permanently on profile with a badge
Sociable award: given upon completing the "about me" profile section, and uploading a profile photo	• +35 points & special icon on user profile page • Achievement showcased permanently on profile with a badge
Eager-to-learn award: user with the most questions (per month)	• +55 points & special icon on user profile page • Achievement showcased permanently on profile with a badge
Popularity award: created an article that was shared over 50 times on social media	• Special icon on user profile page • Achievement showcased permanently on profile with a badge
Best author award: weekly featured article (higher rating)	• +60 points & special icon on user profile page • Article featured in homepage • Achievement showcased permanently on profile with a badge
Bug-hunter award: given to users who reported 5 or more bugs in the platform	• +85 points & special icon on user profile page • Achievement showcased permanently on profile with a badge
Loyalty award: for users that make at least one contribution per day for a month	• +100 points & special icon on user profile page • Achievement showcased permanently on profile with a badge
Financial literacy improvement award: awarded to users who improved their score on the literacy test	• +100 points & special icon on user profile page • Achievement showcased permanently on profile with a badge
Financial educational material award: awarded to users whose educational material received at least 10 ratings and the average rating was above 3/5	• +150 points & special icon on user profile page • Achievement showcased permanently on profile with a badge
Exceptional contribution award: 90% articles/answers rated over 3/5 at a month's period (at least 5 articles/answers)	• +160 points & special icon on user profile page • Achievement showcased permanently on profile with a badge

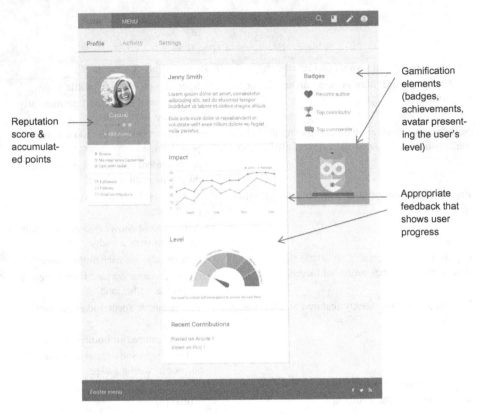

Reputation score & accumulated points

Gamification elements (badges, achievements, avatar presenting the user's level)

Appropriate feedback that shows user progress

Fig. 8. User profile

Monetary Incentives/Awards. Financial incentives that could spark users' interest include the possibility to enter a competition for a tangible reward (e.g. a laptop), which could also act as an initial motivator to join the platform, as well as discounts on financial products provided by local cooperatives and alternative banks. In particular, these tangible rewards will be provided by the members of the "Federation for alternative and ethical banks" (FEBEA) to attract users' participation. These tangible rewards will be provided to users in the form of a competition as an initial motivator to trigger users' interest in the platform, as well as to users that have reached a certain level (experienced and above). This is in line with the studies in [31] that proposed using small monetary rewards as initial motivating factor to attract users, and then sustain their engagement by combining small prizes with other incentive mechanisms, such as gamification elements.

Design of PROFIT User Profile - Dashboard and Social Interaction Activities. Lastly, much attention was also paid in order to provide an appropriate design interface for supporting all aforementioned incentive mechanisms, easily accessible. As can be seen in Fig. 8 which constitutes the mockup of a user's profile, as viewed from that particular user, user level, reputation, and activity points were placed on the top left

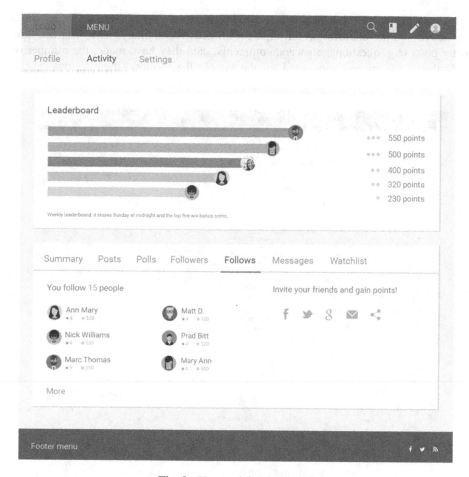

Fig. 9. User activity dashboard

panel, below the user profile photo, in order to make this important information about a user easily accessible. Users can also add some basic profile information and choose whether to showcase or hide their recent activity.

Moreover, visual indicators that provide clear feedback regarding user progress, and can also trigger and/or sustain altruistic motives are going to be included. These consist of an "impact" multi line chart for each user, which shows the number of ratings above 3 her posts receive per week, and a "levels" gauge chart with a dial indicating the current level of the user and the number of points needed to move to the next one. That way, users will have a clear view regarding how much they contribute to improving the financial awareness of other users, as well as their personal progress on the platform. Additionally, they will have the option to make those visualizations private and "hide" them from their – otherwise - public profile pages. Badges which represent achievements on the platform are by default visible to visitors, and below those there is the user avatar – a gamification element that indicates user level in a more "playful" way.

As regards the user dashboard, that would be a private page where a user can access her activity on the platform so far. Apart from activity summary, users can see how many posts (e.g. questions, answers, comments, etc.) they have made, the number of polls they have created and voted to, the people they follow, their followers, etc.

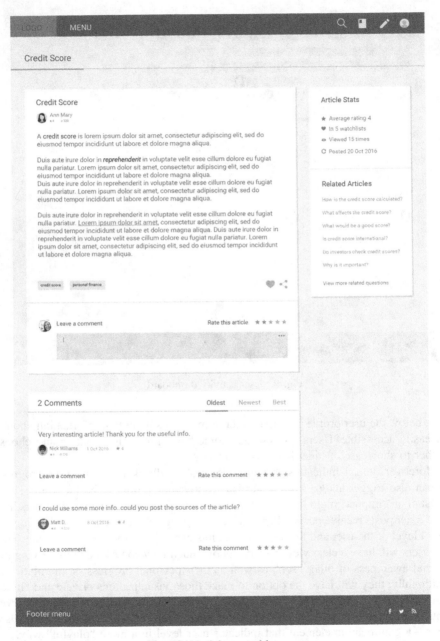

Fig. 10. Article posted by user

Additionally, as mentioned before, users that have reached level 3 (experienced) will also have access to a weekly private leaderboard, showing their ranking among their friends and the points they have accumulated that week.

Last, various social interactions were also taken into consideration and incorporated at the design of the platform. As can also be seen in Fig. 10, users can share posts through their social media accounts, add posts on their watch list to access them later, voice their opinion by rating posts and commenting, and communicate with other users, either by leaving public comments under their posts or via private messages (chat). Users can also see a summary of a user's profile by moving their mouse cursor over her profile picture (Fig. 11) and will be prompted to invite their friends to the platform by gaining points (Fig. 9).

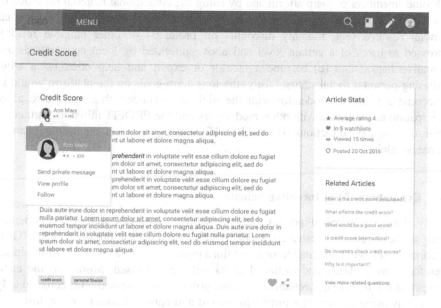

Fig. 11. Social interaction capabilities with the author of the post

3.4   Sustain User Engagement

The PROFIT incentive mechanisms that were described above were designed in a way to sustain user engagement throughout time. It is reasonable that as soon as users reach and safeguard the highest levels or achievements, they might lose their interest in further contributing; this becomes evident from a study by [32], in which users that were given a badge in Wikipedia with no hope of receiving another in the future would later decrease their participation dramatically. In PROFIT, in order to avoid such a situation, we give users the opportunity to gain badges more than once, showing the number each time a badge has been gained next to the badge's icon, while we also incorporate a level "degradation" mechanism, in case users stop contributing, as we explained in Sect. 3.3, and the "Gamification elements" paragraph. More specifically, users need to continue contributing to the platform and provide good quality contributions, in order to sustain their level and respective rights; otherwise they subsequently move down level and lose their rights accordingly. In addition, short-term, mid-term and long-term goals are provided to keep the users' interest alive. More specifically, the short-term goals include: (a) a weekly article award for the highest rated article that is going to be featured on the homepage of the platform, increasing the author's social status on the platform; (b) collection of points in order to level up and unlock privileges in the platform (e.g. become moderator, view advanced activity visualizations); and (c) unlock achievements that can be permanently showcased on user profile in the form of badges, increasing that user's social status.

The medium-term goals consist of the following: (a) the opportunity to increase personal financial knowledge and become more self-accomplished through discussions with other members of the platform and by using the educational material provided on the platform, as well as any informative posts by other users; (b) competition for small financial rewards (e.g. monthly discounts on products) or other tangible rewards addressed to users of a certain level and above provided by local cooperatives and alternative banks; and (c) become a mentor or expert and help other users while increasing personal social status. Lastly, the long-term goals on the platform would be to become a platform moderator with the highest privileges; that could also allow participation in the forum with other moderators and the PROFIT initiators/partners to coordinate the next directions of the PROFIT platform, towards increasing financial awareness globally.

3.5   Evaluation of the Incentive Scheme

Finally, the afore-described incentive scheme, which is currently under development, is going to be iteratively evaluated, using both test data (first prototype of PROFIT platform phase), and real data (second and third prototypes of PROFIT platform phase). Therefore, any usability and technical faults will be addressed. Moreover, the effectiveness of the incentive mechanisms is also going to be assessed and, consequently, the incentive scheme could be further improved and enriched with more sophisticated functionalities. The proof of concept PROFIT platform that will be created will undergo integrated assessment with real users from the pilot organizations and the authors will be in position to: (a) ensure the proper functioning of the proposed reputation system in real-world scenarios; and (b) conduct various research activities allocating users randomly in control and treatment groups, in order to evaluate the effects of various distinct elements of the platform on their engagement levels (e.g. the incorporation of private leaderboards). It should also be noted that interested parties will have the opportunity to experiment with the PROFIT workbench during PROFIT workshops, stands, and demonstrations and provide us with valuable feedback and comments.

4   Conclusions

In this paper we presented the reputation-based incentive mechanism proposed for the PROFIT platform in order to trigger the targeted users' interest, motivate their participation and good quality contributions, and sustain their engagement. In order to design the incentive mechanism for the particular platform, we went through psychological and other studies that investigate users' motives under the context of crowdsourcing, as well as the mapping with the corresponding incentives and incentive mechanisms incorporated in well-known CS platforms, and concluded on using a mix of reputation, gamification, social, extrinsic awards and design elements, appropriately combined in order to appeal to all different target users of the platform. The reputation scheme proposed is neither too simplistic, in order to be capable of deterring malicious or misbehaving activity, nor too complex or time consuming, in order to be efficiently implemented in a real-world platform. The design elements integrated seek to help users understand their unique contribution and impact on the platform, and have a positive effect on their participation level. The effectiveness of the proposed incentive mechanism will be evaluated in real-world scenarios, facilitating us to identify any possible flaws and improve it further.

Acknowledgements.   This work has been supported by the EU HORIZON 2020 project PROFIT (Contract no: 687895).

References

1. Vassileva, J.: Toward social learning environments. IEEE Trans. Learn. Technol. 1(4), 199–214 (2008)
2. McLoughlin, C., Lee, M.J.: Social software and participatory learning: pedagogical choices with technology affordances in the Web 2.0 era. In: ICT: Providing Choices for Learners and Learning, Proceedings Ascilite Singapore 2007, pp. 664–675 (2007)
3. Bryant, T.: Social software in academia. Educause Q. 29(2), 61–64 (2006)
4. Estelles-Arolas, E., Gonzalez-Ladron-De-Guevara, F.: Towards an integrated crowdsourcing definition. J. Inf. Sci. 38, 189–200 (2012)
5. Rouse, A.C.: A preliminary taxonomy of crowdsourcing. In: ACIS 2010 Proceedings, vol. 76, pp. 1–10 (2010)
6. PROFIT Project. http://projectprofit.eu/
7. Leimeister, J.M., Huber, M., Bretschneider, U., Krcmar, H.: Leveraging crowdsourcing: activation-supporting components for IT-based ideas competition. J. Manage. Inf. Syst. 26(1), 197–224 (2009)
8. Brabham, D.C.: Moving the crowd at threadless. Inf. Commun. Soc. 13, 1122–1145 (2010)
9. Quinn, A.J., Bederson, B.B.: Human computation: a survey and taxonomy of a growing field. In: Proceedings of the SIGCHI Conference on Human Factors in Computing Systems. ACM (2011)
10. Hossain, M.: Users' motivation to participate in online crowdsourcing platforms. In: 2012 International Conference on Innovation Management and Technology Research (ICIMTR), pp. 310–315. IEEE (2012)

11. Katmada, A., Satsiou, A., Kompatsiaris, I.: Incentive mechanisms for crowdsourcing platforms. In: Bagnoli, F., Satsiou, A., Stavrakakis, I., Nesi, P., Pacini, G., Welp, Y., Tiropanis, T., DiFranzo, D. (eds.) INSCI 2016. LNCS, vol. 9934, pp. 3–18. Springer, Heidelberg (2016). doi:10.1007/978-3-319-45982-0_1
12. Caps Projects. https://ec.europa.eu/digital-single-market/en/caps-projects
13. DebateHub. http://projects.sigma-orionis.com/catalyst/open-tools/debatehub/
14. Catalyst. http://projects.sigma-orionis.com/catalyst/
15. Wikirate. http://wikirate.eu/
16. Wikirate Interim Report on User Dynamics. http://wikirate.eu/results/deliverables/
17. EnergyUse. https://energyuse.eu/
18. Decarbonet. http://www.decarbonet.eu/
19. CAP4Access. http://www.cap4access.eu/intro.html
20. Vavilis, S., Petković, M., Zannone, N.: A reference model for reputation systems. Decis. Support Syst. **61**, 147–154 (2014)
21. Whitby, A., Jøsang, A., Indulska, J.: Filtering out unfair ratings in bayesian reputation systems. In: Proceedings of 7th International Workshop on Trust in Agent Societies, vol. 6, pp. 106–117 (2004)
22. Jøsang, A., Hird, S., Faccer, E.: Simulating the effect of reputation systems on e-markets. In: Nixon, P., Terzis, S. (eds.) iTrust 2003. LNCS, vol. 2692, pp. 179–194. Springer, Heidelberg (2003). doi:10.1007/3-540-44875-6_13
23. Dellarocas, C.: Immunizing online reputation reporting systems against unfair ratings and discriminatory behavior. In: Proceedings of the 2nd ACM Conference on Electronic Commerce, pp. 150–157. ACM (2000)
24. Deterding, S., et al.: From game design elements to gamefulness: defining gamification. In: Proceedings of the 15th International Academic MindTrek Conference: Envisioning Future Media Environments. ACM (2011)
25. DiTommaso, D.: Beyond gamification: architecting engagement through game design thinking (2011). http://www.slideshare.net/DiTommaso/beyond-gamification-architecting-engagement-through-game-design-thinking. Accessed June 2016
26. Chou, Y.K.: Octalysis – complete gamification framework (2015). http://yukaichou.com/gamification-examples/octalysis-complete-gamification-framework/. Accessed June 2016
27. Brito, J., Vieira, V., Duran, A.: Towards a framework for gamification design on crowdsourcing systems: the GAME approach. In: 2015 12th International Conference on Information Technology-New Generations (ITNG), pp. 445–450. IEEE (2015)
28. Werbach, K., Hunter, D.: For the Win: How Game Thinking Can Revolutionize Your Business. Wharton Digital Press, Philadelphia (2012)
29. Satsiou, A., Tassiulas, L.: Reputation-based resource allocation in P2P systems of rational users. IEEE Trans. Parallel Distrib. Syst. **21**(4), 466–479 (2010)
30. Burke, M., Marlow, C., Lento, T.: Feed me: motivating newcomer contribution in social network sites. In: Proceedings of the SIGCHI Conference on Human Factors in Computing Systems, pp. 945–954. ACM (2009)
31. Massung, E., Coyle, D., Cater, K.F., Jay, M., Preist, C.: Using crowdsourcing to support pro-environmental community activism. In: Proceedings of the SIGCHI Conference on Human Factors in Computing Systems, pp. 371–380. ACM (2013)
32. Restivo, M., van de Rijt, A.: No praise without effort: experimental evidence on how rewards affect Wikipedia's contributor community. Inf. Commun. Soc. **17**(4), 451–462 (2014)
33. PROFIT Deliverable D1.2: Use Cases & User Scenarios, September 2016

Predicting Euro Stock Markets

Ioannis Praggidis[✉], Vasilios Plakandaras, and Eirini Karapistoli

Department of Economics, Democritus University of Thrace,
Panepistimioupoli, 69100 Komotini, Greece
{gpragkid,vplakand}@econ.duth.gr, ikarapis@duth.gr
http://www.econ.duth.gr/index.en.shtml

Abstract. Forecasting exercises are mostly concentrated on the point estimation of future realizations of stock returns. In this paper we try to forecast the direction of the Eurostoxx 50. Under a Dynamic Probit framework we test whether subsequent sign reversals can be accurately forecasted. To this end, we make use of industrial portfolios constructed in the spirit of Fama and French. Furthermore, we augment the forecasting models with macroeconomic variables. Finally, we construct a new sentiment index based on the news for Oil prices. Results show, that the out-of-sample forecasting accuracy approximates 80%.

Keywords: Eurostoxx 50 · Portfolio industries · Dynamic probit models · Uncertainty

1 Introduction

The efficient market hypothesis Fama [9] postulates that publicly available information cannot be used to predict stock returns. Nevertheless, there are numerous empirical studies that reach to a contradicting outcome. Among others Chen [6], Nyberg [19,20], Driesprong et al. [8] and Hong et al. [14] provide significant evidence that macroeconomic variables, industry portfolios and oil prices can predict future movements in stock markets. Nevertheless, the vast majority of the existing literature studies the U.S. stock market, providing little empirical results to forecasting European stock indexes.

In this paper we contribute to the recent literature by examining the ability to forecast the Eurostoxx 50 index. Interestingly Eurostoxx 50 is an under investigated index although refers to the financial conditions in the Eurozone. In order to do so we first construct the 38 industrial portfolio using almost 4500 stocks traded in European stock exchange markets in the vein of Fama and French [10]. According to Hong et al. [14], the gradual information diffusion hypothesis explains why information extracted from specific sectors can act as leading indicators for the market index. The basic idea is that certain investors, such as those that specialize in trading the broad market index, receive information originating from particular industries such as commercial real estate or commodities like metals only with a lag. In contrast, investors that specialize in a specific market cannot detect the trends of the entire stock market and

© Springer International Publishing AG 2016
A. Satsiou et al. (Eds.): IFIN and ISEM 2016, LNCS 10078, pp. 81–97, 2016.
DOI: 10.1007/978-3-319-50237-3_3

thus are unable to transmit efficiently their specialized knowledge of the sector to the entire market. Combining the information from different sources (industries) we expect to be able to grasp their informational content in forecasting the Eurostoxx 50 index.

In order to measure the forecasting ability of our models to the market efficiency assumption, we use the Random Walk model as a benchmark. The use of the RW model as a forecasting benchmark is also a contemporaneous testing of the Efficient Market Hypothesis (EMH). Proposed by Eugene Fama [9], states that the determination of prices in an efficient market follows a random walk and thus it is impossible to create a forecasting model that achieves sustainable positive returns on the long-run. The EMH is usually presented in three forms; the weak, the semi-strong and the strong form of efficiency. We have a weak-form efficient market when historic prices of the variable in question cannot forecast the future ones, as the generating mechanism of prices follows a RW. Thus, autoregressive models, in these cases, have no forecasting power and the best forecast about next period's price is today's price. Semi-strong efficiency imposes more strict assumptions in that all historic prices and all publicly available information is already reflected in current asset prices and thus they cannot be used successfully in forecasting. Finally, the strong EMH builds on the semi-strong case adding all private information and thus making impossible to forecast successfully the future evolution of an asset's price.

Furthermore, Driesprong *et al.* [8] points that changes in oil prices predict stock market returns worldwide finding significant predictability in both developed and emerging markets. Following Driesprong *et al.* [8] we use a new developed proxy for sentiment in oil market. We test the predictability of a sentiment index as sentiment measure acts as a leading indicator for oil prices [24].

The remainder of this paper is organized as follows. In Sect. 2, we outline the recent advances in the area of market sentiment analysis. In Sect. 3, we present the methodology we followed in our market predictive sentiment analysis and in Sect. 4, we describe the textual and market data that are used as input in our system. In Sect. 5, we provide a detailed description of the probit model and its specifications, while empirical results from our analysis are provided in Sect. 6. Finally, Sect. 7 provides concluding remarks and points towards future research avenues.

2   News Media Sentiment and the Stock Market

The recent years, there has been a significant amount of work on using textual resources from the Web (such as financial news articles, online reviews, blogs, Twitter, etc.) to predict the stock market as well as financial and economic variables of interest [27]. Much of this work has relied on some form of *sentiment analysis* to represent the text [17]. Sentiment analysis targets open issues in various fields (including politics, psychology, finance, and society), because sentiments' understanding can largely impact interactions, policies, and decision-making. Due to its importance, this form of textual information processing has become a growing part of the empirical finance research.

Typically, the two major sentiment analysis tasks studied today are *subjectivity* detection and *polarity* detection. The most common approach to deal with these tasks is either to train a machine learning classifier on a labeled corpus (supervised learning) and apply the learned model on the desired test set, or use a predefined dictionary consisting of words that are annotated with their semantic orientation value (polarity and strength). The lexicon-based approaches are based on the assumption that the polarity of a given text is the sum of the semantic orientation of each word or phrase contained in it (unsupervised learning). Limited work has also been conducted on hybrid methodologies as well as on ontology-supported approaches.

2.1 Lexicon-Based Approaches

The lexicon-based approach is an unsupervised technique that extracts the sentiment of a text using lexicons (dictionaries) that may be either domain-specific (such as finance, politics, psychology, etc.) or domain-independent [29]. Such dictionaries consist of words that are annotated with their semantic orientation value (polarity and strength); usually a positive or negative value is assigned to each word (for example, the score for the word "good" is 0.65 and the score for the word "bad" is 0.85).

Dictionaries for lexicon-based approaches can be created either manually [25, 28], or automatically, using seed words to expand the list of words [13, 29, 30]. Much of the lexicon-based research has focused on using adjectives as indicators regarding the semantic orientation of the text [13, 15, 26, 31]. According to this, a list of adjectives and corresponding semantic orientation values is compiled into a dictionary. Then, for any given text, all adjectives are extracted and annotated according to their semantic orientation value using the dictionary scores. The semantic orientation scores are in turn aggregated into a single score for the text. However, lexicon-based approaches suffer from their absolute dependence on lexicons, which are often characterized by word shortage or inappropriate assignment of semantic orientation values [18].

2.2 Machine Learning Approaches

The text classification approach is essentially a supervised classification task, which involves building classifiers from labelled instances of texts or sentences. The introduction of the machine learning approach in sentiment analysis originates from Bo Pang *et al.* [21]. This approach is also referred to as a statistical or machine learning approach. The majority of the statistical text classification research relies on Nave Bayes, Support Vector Machines (SVM), and Max entropy classifiers trained on a particular data set using features such as unigrams or bigrams, and with or without part-of-speech (POS) labels [21, 23].

Classifiers built using the supervised methods may achieve high accuracy when detecting the polarity of a text [1, 3, 4]. However, although such classifiers perform very well in the domain that they are trained on, their performance degrades when the same classifier is used in a different domain [2]. Moreover, utilizing an individual classifier might result in poor sentiment detection, since the performance of each classifier varies significantly, when for instance someone is using different features or weight measures. Therefore, for overcoming the deficiencies of each classifier and to proceed with a more robust and successful sentiment detection process, ensemble classifiers are created, which represent a combination of multiple classifiers Xia et al. [32].

2.3 Hybrid Approaches

It became apparent that the lexicon-based approaches suffer from their absolute dependence on lexicons, which are often characterized by word shortage or inappropriate assignment of semantic orientation values. While, machine-learning approaches overcome these limitations, their need for a large volume of training data also lessens their advantage [12].

Accordingly, the hybrid approach targets at solving these limitations by combining the aforementioned approaches. An exemplar use case scenario for a hybrid approach is to follow a two-step process; i.e., to initially generate a set of training data by automatically identifying the texts' semantic orientation score (using a lexicon-based approach), and then to proceed with the classification (using a machine learning approach) that is independent from the lexicons' limitations [11]. As Prabowo and Thelwall [22] point out, hybrid approaches enhance the stability and accuracy of existing methodologies while exploring the strong characteristics of each of them.

2.4 Ontology-Based Approaches

Recently, semantic web-based sentiment analysis that relies on the usage of ontologies has started to attract attention. Ontologies are not a classification method in itself, but they can be used to enhance existing classification methods. Zhou and Chaovalit [33] were among the first to use ontologies for this purpose. They incorporated an ontology to a supervised machine learning technique and a basic term counting method. In their work, as in most other works, the ontology contained semantic features and was used only to identify these features inside the text. Cadilhac et al. [5] on the other hand, they did not use the ontology only as a taxonomy by taking into account the "is a relation" between concepts. Instead, they presented an approach that uses other relations between concepts for creating summaries of opinions and computing the SO scores.

2.5 Our Approach

In this paper, we tackle the problem of polarity detection in financial news headlines. Using a simple Java client API (with no third party libraries), we query the Virtuoso SPARQL endpoint of PoolParty[1]; a Thesaurus Management Tool (TMT) for the Semantic Web, and fetch 14205 date-stamped news headlines

---

[1] https://www.poolparty.biz.

related to the Eurostoxx 50 index. We then create two different representations for the gathered textual resources (one based on machine learning and one on lexicon-based learning) and train one classifier for each one of them.

3   Methodology

In the current work, the relevant market-predictive sentiment analysis process is considered to have four major phases, namely: information retrieval, pre-processing, model creation and prediction, and evaluation (see Fig. 1). Information retrieval is the activity of obtaining information resources relevant to a financial instrument from a collection of online information sources. Our search was based on full-text indexing and was restricted to the digital archive of the investing.com website. All those articles containing one of the following terms: 'oil prices', 'crude oil', 'brent', 'future prices', 'WTI' or 'OPEC' (including their variants) were stored in PoolParty.

After acquiring the textual resources, we process the input data so that it can be fed into our models. Natural language processing (NLP) of document texts includes among others part-of-speech (POS) tagging, chunking and named entity recognition. We can also apply tokenisation, sentence splitting, and morphological analysis in order to further clean the text [7]. Having transformed the unstructured text into a representative format that is structured, it can now be processed by the machine. Two model representations of the news headlines are created (one model based on text-based learning and one on lexicon-based learning). For the text-based representation, we used the binary, term frequency (tf) and $tf \cdot idf$ n-gram models. We set $n = 1, 2$ resulting into six representations in total. Stop-word removal and stemming were ignored. For the lexicon-based representation, we used the Harvard dictionary[2]. Instead of assigning the majority class label on each news headline, we counted the sum of nouns, verbs, adjectives, and adverbs as indicated by [15]. In the presence of negation, the polarity value of the term was inverted. We used these four features to learn the lexicon-based model, and we compare our results with the simple counting method of the lexicon itself.

Regarding the model creation and prediction phase, the proposed system is using Multinomial Naive Bayes (MNB) and Support Vector Machines (SVM) as provided by Weka[3]. The MNB model is a popular method for text classification due to its computational efficiency and relatively good predictive performance. MNB computes the posterior probability of a class, based on the multinomial distribution of the words in the document. The model works with the bag-of-words (BOWs) feature extraction which ignores the position of the word in the document. Finally, it uses Bayes Theorem to predict the probability that a given feature set belongs to a particular label. On the other hand, the main principle of SVMs is to determine linear separators in the search space which can best separate the different classes. Text data are ideally suited for SVM classification

---

[2] http://www.wjh.harvard.edu/~inquirer/.
[3] http://www.cs.waikato.ac.nz/ml/weka/.

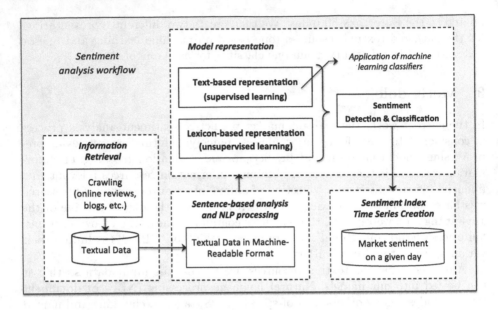

Fig. 1. The proposed market-predictive sentiment analysis process.

because of the sparse nature of text, in which few features are irrelevant, but they tend to be correlated with one another and generally organized into linearly separable categories. We chose MNB to be applied on the text-based representations and SVM for the lexicon-based representation due to its capability of dealing with double-valued attributes.

For the evaluation, we randomized the test dataset and used a 70%–30% split for the purpose of training and evaluation respectively. Table 1 presents the accuracy for all examined text-based representation models. As expected, bigrams outperformed unigrams. Several differences were also observed moving from binary to the $tf \cdot idf$ weighting scheme. Table 2 illustrates the results obtained by SVM compared to the "counting" method using the Harvard dictionary alone, revealing the superiority of the SVM approach. Our findings are also consistent with the claims supporting that adjectives carry more sentimental weight.

Figure 2 illustrates the monthly aggregated sentiment polarity index. First, we extract the daily market sentiment polarity index with respect to a specific financial instrument or indicator (the oil prices in our case) based on the sentiment analyses of news headlines obtained during the last 24 h. The aggregation of the sentiment to the month level is done by aggregating the sentiment of the daily textual resources that were tagged positive and negative in the previous step. Our empirical results show that the MNB model produces statistically significant results, while the lexicon based method does not produce any significant results. Granger causality test also indicates a statistically significant causation stemming from oil sentiment index to oil prices while the opposite is

Table 1. Accuracy achieved for the text-based model

Representation	Binary		Term freq.		tf-idf	
ngrams	1	2	1	2	1	2
Accuracy (%)	74.89	**80.06**	75.31	**81.92**	76.87	**82.24**

Table 2. Accuracy achieved by SVM applied on the lexicon-based model

	Adverb	Noun	Verb	Adjective
Accuracy (%)	53.4	54.9	58.8	**66.93**

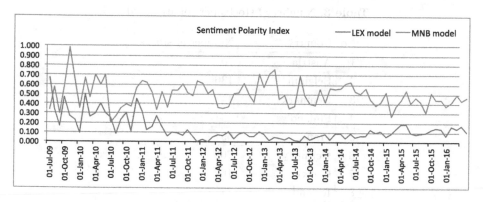

Fig. 2. Monthly aggregated sentiment polarity index.

rejected. Overall, we find that the sentiment is an important consideration when explaining crude oil prices using data from online textual resources.

4   Datasets

The propose system is taking two sources of information as input, namely, *textual data* from online resources (financial websites), and *market data*.

4.1   Textual Data

The study covers the period 21 July 2009 to 06 May 2016. Our sentiment analysis relies on daily content from `investing.com`; a leading financial platform offering real-time quotes, streaming charts, financial news, technical analysis and more. We searched the digital archive of `investing.com` to obtain articles containing one of the following terms: 'oil prices', 'crude oil', 'brent' or 'future prices' (including their variants). In other words, for an article to meet our criteria, it must contain at least one term pertaining to the Eurostoxx index. The final sample contains 14205 news headlines together with their associated body text, which are fetched by querying the Virtuoso SPARQL endpoint of PoolParty using a simple web crawling algorithm. The sentiment polarity index is published each day and is calculated automatically based on the sentiment analyses of news headlines obtained during the last 24 h.

4.2 Market Data

In order to construct the industry portfolios, we compile 4442 daily stock prices from Yahoo finance! for the period 2008 to 2016, traded in 15 European countries. The countries and their number of stocks are reported in Table 3[4].

Table 3. Number of stocks per country used

No	Country	Number of stocks
1	Austria	70
2	Belgium	136
3	Denmark	148
4	Finland	131
5	France	697
6	Germany	703
7	Greece	167
8	Iceland	17
9	Italy	293
10	Netherlands	98
11	Norway	175
12	Portugal	45
13	Spain	144
14	Turkey	404
15	United Kingdom	1210
		Total: 4442

We take first logarithmic differences in order to compute daily returns, and we convert daily to monthly returns using monthly averages. The selection of the monthly forecasting horizon of the Eurostoxx 50 index is based on data availability. Following the classification system of Fama and French [10] in 38 industry portfolios, we construct 38 equally weighted industry portfolios, depicted in Table 4. We end up with 20 monthly portfolios, according to the activity of each listed company and 99 monthly observations.

As stated in the Introduction section, the scope of this exercise is to forecast the future direction of the Eurostoxx 50 index. After compiling daily values

---

[4] The selected countries and the number of stocks are upon availability of the data.

Table 4. Industry portfolios

	Industry code	Industry description (SIC codes)	Number of stocks
1	Agric	Agriculture, forestry, and fishing (0100–0999)	31
2	Mine	Mining (1000–1299)	74
3	Oil	Oil and Gas Extraction (1300–1399)	102
4	Stone	Non-metalic Minerals Except Fuels (1400–1499)	0
5	Cnstr	Construction (1500–1799)	129
6	Food	Food and Kindred Products (2000–2099)	156
7	Smoke	Tobacco Products (2100–2199)	3
8	Txtls	Textile Mill Products (2200–2299)	0
9	Apprl	Apparel and other Textile Products (2300–2399)	8
10	Wood	Wood Lumber and Wood Products (2400–2499)	0
11	Chair	Furniture and Fixtures (2500–2599)	33
12	Paper	Paper and Allied Products (2600–2661)	27
13	Print	Printing and Publishing (2700–2799)	86
14	Chems	Chemicals and Allied Products (2800–2899)	261
15	Ptrlm	Petroleum and Coal Products (2900–2999)	0
16	Rubbr	Rubber and Miscellaneous Plastics Products (3000–3099)	5
17	Lethr	Leather and Leather Products (3100–3199)	0
18	Glass	Stone, Clay and Glass Products (3200–3299)	0
19	Metal	Primary Metal Industries (3300–3399)	36
20	MtlPr	Fabricated Metal Products (3400–3499)	0
21	Machn	Machinery, Except Electrical (3500–3599)	189
22	Elctr	Electrical and Electronic Equipment (3600–3699)	307
23	Cars	Transportation Equipment (3700–3799)	66
24	Instr	Instruments and Related Products (3800–3879)	38
25	Manuf	Miscellaneous Manufacturing Industries (3900–3999)	0
26	Trans	Transportation (4000–4799)	81
27	Phone	Telephone and Telegraph Communication (4800–4829)	139
28	TV	Radio and Television Broadcasting (4830–4899)	32
29	Utils	Electric, Gas, and Water Supply (4900–4949)	167
30	Garbg	Sanitary Services (4950–4959)	33
31	Steam	Steam Supply (4960–4969)	0
32	Water	Irrigation Systems (4970–4979)	0
33	Whlsl	Wholesale (5000–5199)	227
34	Retail	Retail Stores (5200–5999)	376
35	Money	Finance, Insurance, and Real Estate (6000–6999)	894
36	Srvc	Services (7000–8999)	885
37	Govt	Public Administration (9000–9999)	32
38	Other	Everything else	8

of the index from Yahoo finance!, we take monthly averages of the index. The classification of the stock market in bear and bull market is achieved directly by comparing the successive values of the index, constructing a binary index of ones and zeros, to be fed in the forecasting model. Unlike Nyberg [20], we do not use filer algorithms in order to discern bear and bull markets, since this category of filters tends to detect prolonged periods of market states (15 months at minimum) that cannot be used effectively in active index trading.

5   Probit Specification

5.1   The Model

The majority of previous empirical studies that considers different regimes in stock market exploits Markov switching models. Although the specific category of models has the advantage of providing state-dependent inferences, the main drawback is that it is based on an unobservable Markov switching process that cannot be explicitly described. In our study we consider a binary response model that predicts the future direction of the index as a binary time series. Considering the binary state variable s_t, the value 1 states a bull and 0 a bear market as follows:

$$s_t = \begin{cases} 1 & \textit{bull} \text{ market at time t,} \\ 2 & \textit{bear} \text{ market at time t,} \end{cases} \tag{1}$$

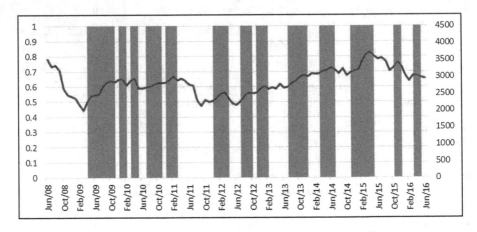

Fig. 3. Bear vs bullish periods of the Eurostoxx 50 index. Note: the continuous line reports actual values of the index, while the shaded areas are bullish periods.

for $t = 1, 2, 3, .., T$ the range of the monthly observations. Denoting the conditional expectation $E_{t-1}(s_t|\Omega_{t-1})$ in the information set Ω_{t-1} at time $t-1$, the conditional probability at time t that the market is in a bull state is:

$$p_t = E_{t-1}(s_t|\Omega_{t-1}) = P_{t-1}(s_t = 1) = \Phi(\pi_t) \tag{2}$$

Where π_t is a linear combination of variables and $\Phi(\bullet)$ is the normal cumulative distribution function. Naturally, the conditional probability of the bull market is the complement of the bear market probability $P_{t-1}(s_t = 0) = 1 - p_t$. In order to predict the linear function π_t we consider different static and dynamic models. Starting from the univariate probit model [6]

$$\pi_t = \omega + \chi'_{t-h}\beta \tag{3}$$

where ω is a constant, β is the coefficients vector and χ'_{t-h} a matrix of predictive regressors. The index h denotes the forecasting horizon. The popular static model is extended by adding lags of the dependent variable s_t resulting in the autoregressive static model

$$\pi_t = \omega + \alpha(s_{t-1})\chi'_{t-h}\beta \tag{4}$$

or by adding lags of the dependent variable π_t, leading to the dynamic model proposed by Kauppi and Saikkonen [16]:

$$\pi_t = \omega + \delta\pi_{t-1} + \chi'_{t-h}\beta \tag{5}$$

By recursive substitution, Eq. 5 can be seen as an infinite order static Eq. 4 where the whole history of the values of the predictive variables included has an effect χ_{t-h} on the conditional probability. Thus, if the longer history of explanatory variables included in χ_{t-h} are useful to predict the future market status, the autoregressive Eq. 5 may offer a parsimonious way to specify the predictive model. A natural extension would be the dynamic autoregressive model:

$$\pi_t = \omega + \alpha(s_{t-1}) + \delta\pi_{t-1} + \chi'_{t-h}\beta \tag{6}$$

6 Preliminary Results

6.1 Empirical Results

We forecast the future direction of the Eurostoxx 50 for 1 to 12 months ahead based on the static models Eqs. 3 and 4 and the dynamic models Eqs. 5 and 6. After forecasting the value of π_t we compute the probability $\Phi(\pi_t)$ from the normal cumulative distribution function. We assign a threshold in the probability of belonging to a bull or bear market. A popular selection is 0.5, but as we observe from Fig. 3, there are more bear than bullish periods. Since the mean value of s_t is 0.3544 we also compute bear and bull periods according to the threshold 0.35. Each time we keep the last 18 observations from the training phase of the model in order to measure the out-of-sample forecasting accuracy.

As a first step, we forecast π_t considering only the portfolio of one industry at a time as a regressor, in order to observe the potential contribution of each industry in forecasting the market. In Table 5, we report the in and out-of sample forecasting accuracy for the most accurate models. The selection of the

Table 5. Results of the estimated four models

	Static				AR static				Dynamic				AR dynamic			
h	R^2	In	Out	Sector	R^2	In	Out	Sector	R^2	In	Out	Sector	R^2	In	Out	Sector
1	0.22	0.73	0.78	TV	0.22	0.83	0.78	Smoke	0.24	0.72	0.72	TV	0.29	0.74	0.61	TV
2	0.00	0.65	0.61	Smoke	0.13	0.70	0.61	Apprl	0.01	0.64	0.61	Oil	0.19	0.70	0.56	Apprl
3	0.00	0.66	0.67	Phone	0.08	0.72	0.67	Phone	0.00	0.66	0.67	Phone	0.14	0.71	0.67	Cnstr
4	−0.01	0.68	0.67	Garbg	0.09	0.72	0.72	Smoke	0.02	0.67	0.50	Paper	0.17	0.71	0.56	Garbg
5	−0.02	0.62	0.67	Rubbr	0.08	0.70	0.67	Garbg	0.05	0.66	0.44	Utils	0.16	0.74	0.67	Rubbr
6	−0.01	0.64	0.72	Garbg	0.08	0.71	0.67	Smoke	0.04	0.64	0.56	Agric	0.14	0.71	0.61	Smoke
7	0.02	0.68	0.78	TV	0.10	0.71	0.78	TV	0.05	0.63	0.50	Agric	0.15	0.72	0.61	Apprl
8	0.01	0.65	0.61	Agric	0.07	0.72	0.72	Agric	0.04	0.61	0.50	Agric	0.13	0.69	0.56	Garbg
9	0.01	0.63	0.39	Oil	0.07	0.70	0.72	Rubbr	0.03	0.59	0.44	Agric	0.15	0.67	0.67	Smoke
10	0.04	0.59	0.56	Agric	0.09	0.68	0.72	Agric	0.04	0.61	0.50	Agric	0.15	0.70	0.67	Govt
11	0.02	0.63	0.56	Agric	0.08	0.68	0.67	Smoke	0.02	0.63	0.56	Agric	0.13	0.71	0.67	Agric
12	0.00	0.61	0.67	Instr	0.07	0.72	0.72	Instr	0.00	0.60	0.67	Instr	0.13	0.72	0.67	TV
	Static				AR static				Dynamic				AR dynamic			
h	R2	In	Out	Sector	R2	In	Out	Sector	R2	In	Out	Sector	R2	In	Out	Sector
1	0.22	0.69	0.72	TV	0.22	0.73	0.78	Smoke	0.24	0.71	0.72	TV	0.29	0.76	0.72	TV
2	0.00	0.52	0.61	Smoke	0.13	0.68	0.61	Apprl	0.01	0.52	0.50	Oil	0.19	0.71	0.67	Apprl
3	0.00	0.50	0.39	Phone	0.08	0.72	0.72	Phone	0.00	0.46	0.39	Phone	0.14	0.72	0.67	Cnstr
4	−0.01	0.47	0.28	Garbg	0.09	0.71	0.72	Smoke	0.02	0.57	0.28	Paper	0.17	0.72	0.61	Garbg
5	−0.02	0.51	0.39	Rubbr	0.08	0.72	0.72	Garbg	0.05	0.51	0.28	Utils	0.16	0.73	0.67	Rubbr
6	−0.01	0.60	0.44	Garbg	0.08	0.67	0.67	Smoke	0.04	0.52	0.28	Agric	0.14	0.67	0.67	Smoke
7	0.02	0.50	0.44	TV	0.10	0.74	0.72	TV	0.05	0.54	0.28	Agric	0.15	0.71	0.67	Apprl
8	0.01	0.54	0.39	Agric	0.07	0.70	0.67	Agric	0.04	0.62	0.28	Agric	0.13	0.66	0.56	Garbg
9	0.01	0.53	0.33	Oil	0.07	0.64	0.72	Rubbr	0.03	0.59	0.28	Agric	0.15	0.73	0.67	Smoke
10	0.04	0.61	0.44	Agric	0.09	0.64	0.61	Agric	0.04	0.61	0.39	Agric	0.15	0.64	0.61	Govt
11	0.02	0.57	0.44	Agric	0.08	0.68	0.67	Smoke	0.02	0.54	0.33	Agric	0.13	0.71	0.67	Agric
12	0.00	0.51	0.44	Instr	0.07	0.70	0.72	Instr	0.00	0.54	0.33	Instr	0.13	0.67	0.67	TV

Table 6. Best fitted results for each forecasting period

	1	2	3	4	5	6	7	8	9	10	11	12
In-sample accuracy												
RW	0.73	0.53	0.49	0.49	0.50	0.49	0.53	0.54	0.49	0.49	0.48	0.55
Model	0.83	0.71	0.72	0.72	0.74	0.67	0.71	0.72	0.70	0.68	0.71	0.72
Out-of-sample accuracy												
RW	0.71	0.67	0.60	0.50	0.62	0.42	0.60	0.60	0.56	0.50	0.29	0.50
Model	0.78	0.67	0.72	0.72	0.67	0.67	0.78	0.72	0.72	0.72	0.67	0.72
Sector	Smoke	Apprl	Phone	Smoke	Rubbr	Smoke	TV	Agric	Rubbr	Agric	Agric	Instr

most accurate model among the available portfolios is based on the maximum McFadden R square of the training models.

In Table 6, we depict only the best results for each forecasting horizon, as well as the forecasting accuracy of the benchmark RW model. As we observe from Table 6, the considered forecasting models exhibit similar or outperform by large the RW in all forecasting horizons when it comes to out-of-sample accuracy. Another interesting conclusion is that the agricultural industry outperforms all other sectors in the longer forecasting horizons. Thus, we reject even the weak

Table 7. Results from all 20 available industry portfolios

	Static		AR static		Dynamic		AR dynamic	
In sample								
h	In	Out	In	Out	In	Out	In	Out
1	1.00	0.78	1.00	0.67	1.00	0.67	1.00	0.67
2	0.79	0.44	0.82	0.61	0.91	0.50	0.88	0.50
3	0.79	0.61	0.82	0.56	0.89	0.67	0.92	0.61
4	0.91	0.33	0.93	0.50	0.93	0.44	1.00	0.50
5	0.77	0.44	0.96	0.44	1.00	0.44	0.97	0.39
6	0.89	0.44	1.00	0.50	0.99	0.44	1.00	0.44
7	0.79	0.56	0.82	0.61	0.90	0.56	0.92	0.56
8	0.83	0.44	0.83	0.50	0.94	0.50	0.99	0.39
9	0.89	0.50	0.96	0.56	0.90	0.50	1.00	0.56
10	0.81	0.33	0.84	0.39	0.94	0.44	0.91	0.39
11	0.88	0.56	0.88	0.56	0.88	0.44	0.88	0.61
12	0.97	0.50	0.99	0.61	0.94	0.39	0.94	0.50
Out of sample								
h	In	Out	In	Out	In	Out	In	Out
1	1.00	0.78	1.00	0.67	1.00	0.67	1.00	0.67
2	0.74	0.39	0.81	0.61	0.91	0.39	0.88	0.50
3	0.76	0.56	0.79	0.56	0.87	0.67	0.91	0.61
4	0.92	0.39	0.93	0.50	0.92	0.44	1.00	0.50
5	0.76	0.44	0.92	0.44	1.00	0.44	0.96	0.39
6	0.82	0.44	1.00	0.50	0.97	0.44	1.00	0.44
7	0.75	0.61	0.82	0.61	0.89	0.56	0.93	0.56
8	0.79	0.44	0.82	0.50	0.92	0.50	0.96	0.39
9	0.91	0.50	0.96	0.56	0.91	0.50	1.00	0.56
10	0.87	0.44	0.84	0.50	0.96	0.39	0.91	0.33
11	0.84	0.50	0.85	0.56	0.82	0.39	0.88	0.56
12	0.91	0.50	0.99	0.61	0.94	0.39	0.90	0.50

form of market efficiency. In Table 7, we report the results from training models using all the available 20 industry portfolios as regressors.

In Table 8, we concentrate on the forecasting accuracy of the most accurate model per forecasting horizon, vis a vis with the RW model. As it can be seen from this Table, the forecasting ability of the forecasting models that incorporate the entire dataset is smaller than the use of specific industry portfolios. More specifically, when we focus on the out-of-sample accuracy, the accuracy of the models is smaller than the random selection (random guess) of the future

Table 8. Forecasting accuracy of the most accurate model per forecasting horizon.

	1	2	3	4	5	6	7	8	9	10	11	12
In-sample accuracy												
RW	0.73	0.53	0.49	0.49	0.50	0.49	0.53	0.54	0.49	0.49	0.48	0.55
Model	1.00	0.82	0.89	0.93	0.94	0.96	0.82	0.94	0.96	0.94	0.88	0.99
Out-of-sample accuracy												
RW	0.71	0.67	0.60	0.50	0.62	0.42	0.60	0.60	0.56	0.50	0.29	0.50
Model	0.78	0.61	0.67	0.5	0.50	0.56	0.61	0.50	0.56	0.44	0.61	0.61

Table 9. Forecasting power of oil sentiment index

Oil sentiment index			
2010M10–2016M03			
Horizon	In	Out	Model
1	0.68	0.55	AR static
2	0.61	0.60	AR dynamic
3	0.60	0.55	AR static
4	0.60	0.55	AR static
5	0.66	0.60	AR dynamic
6	0.60	0.55	AR static
7	0.67	0.60	AR dynamic
8	0.68	0.60	AR dynamic
9	0.70	0.60	AR dynamic
10	0.69	0.60	AR dynamic
11	0.71	0.60	AR dynamic
12	0.65	0.60	AR dynamic

direction of the market. This fact could be attributed to the introduction of information to the model that is irrelevant to forecast (noise).

6.2   Oil's Sentiment Predictive Power

In this section we present the results of the oil sentiment index as a predictor of the Eurostoxx 50. As previously we estimate four different Probit models and in Table 9 we present the results. We present here results of the oil sentiment index produced by using Multinomial Naive Bayes and SVM system. The other two methods do not produce any statistically significant results,

Results above show that the oil sentiment index has predictive power on the Eurostoxx 50. It is interesting that almost for any estimated forecast period the AR Dynamic model produces the best results. Using the oil sentiment index we can forecast in 60% accuracy the future sign reversals of Eurostoxx 50.

7   Conclusions

In this paper, we forecast the future direction of the Eurostoxx 50 index using monthly data for various horizons. In doing so, we develop static and dynamic probit models, while we construct industry portfolios from a plethora of stocks traded in different stock markets. Our empirical findings suggest that the dynamic models do not improve the forecasting ability of the models upon the use of static version of the model, while certain sectors of the stock market outperform the RW model, leading to profitable trading strategies. In future work, we intend to expand our sample to 2004 for both stock prices and oil sentiment. It is also crucial to estimate the predictive power of specific macroeconomic variables as well as combinations of industry portfolios and macroeconomic variables. Another interesting aspect will be to estimate whether industry portfolios from the core countries of the Eurozone have different predictive power than industry portfolios from the periphery countries. With regard to sentiment analysis, we intend to apply several accuracy-boosting machine learning techniques to the sentiment classification problem.

Acknowledgements. This work has been supported by the EU HORIZON 2020 project PROFIT (Contract no: 687895).

References

1. Alistair, K., Diana, I.: Sentiment classification of movie and product reviews using contextual valence shifters. In: Proceedings of FINEXIN (2005)
2. Aue, A., Gamon, M.: Customizing sentiment classifiers to new domains: a case study. In: Proceedings of Recent Advances in Natural Language Processing (RANLP), pp. 1–2 (2005)
3. Bartlett, J., Albright, R.: Coming to a theater near you! Sentiment classification techniques using SAS text miner. In: SAS Global Forum 2008 (2008)
4. Boiy, E., Hens, P., Deschacht, K., Moens, M.F.: Automatic sentiment analysis in on-line text. In: ELPUB, pp. 349–360 (2007)
5. Cadilhac, A., Benamara, F., Aussenac-Gilles, N.: Ontolexical resources for feature based opinion mining: a case-study. In: 23rd International Conference on Computational Linguistics, p. 77. Citeseer (2010)
6. Chen, S.S.: Predicting the bear stock market: macroeconomic variables as leading indicators. J. Bank. Finance **33**(2), 211–223 (2009)
7. Cunningham, H., Maynard, D., Bontcheva, K., Tablan, V.: Gate: a framework and graphical development environment for robust NLP tools and applications. In: Proceedings of 40th Anniversary Meeting of the Association for Computational Linguistics (ACL) (2002)
8. Driesprong, G., Jacobsen, B., Maat, B.: Striking oil: another puzzle? J. Financ. Econ. **89**(2), 307–327 (2008)
9. Fama, E.: The behavior of stock-market prices. J. Bus. **38**(1), 34–105 (1965)
10. Fama, E., French, K.: Industry costs of equity. J. Financ. Econ. **43**(2), 153–193 (1997)

11. Ghiassi, M., Skinner, J., Zimbra, D.: Twitter brand sentiment analysis: a hybrid system using n-gram analysis and dynamic artificial neural network. Expert Syst. Appl. **40**(16), 6266–6282 (2013)
12. Gonçalves, P., Araújo, M., Benevenuto, F., Cha, M.: Comparing and combining sentiment analysis methods. In: Proceedings of the first ACM Conference on Online Social Networks, pp. 27–38. ACM (2013)
13. Hatzivassiloglou, V., McKeown, K.R.: Predicting the semantic orientation of adjectives. In: Proceedings of the Eighth Conference on European Chapter of the Association for Computational Linguistics, pp. 174–181. Association for Computational Linguistics (1997)
14. Hong, H., Torous, W., Valkanov, R.: Do industries lead stock markets? J. Financ. Econ. **83**(2), 367–396 (2007)
15. Hu, M., Liu, B.: Mining and summarizing customer reviews. In: Proceedings of the Tenth ACM SIGKDD International Conference on Knowledge Discovery and Data Mining, pp. 168–177. ACM (2004)
16. Kauppi, H., Saikkonen, P.: Predicting us recessions with dynamic binary response models. Rev. Econ. Stat. **90**(4), 777–791 (2008)
17. Liu, B.: Sentiment analysis and opinion mining. Synth. Lect. Hum. Lang. Technol. **5**(1), 1–167 (2012)
18. Nassirtoussi, A.K., Aghabozorgi, S., Wah, T.Y., Ngo, D.C.L.: Text mining for market prediction: a systematic review. Expert Syst. Appl. **41**(16), 7653–7670 (2014)
19. Nyberg, H.: Forecasting the direction of the us stock market with dynamic binary probit models. Int. J. Forecast. **27**(2), 561–578 (2011)
20. Nyberg, H.: Predicting bear and bull stock markets with dynamic binary time series models. J. Bank. Finance **37**(9), 3351–3363 (2013)
21. Pang, B., Lee, L., Vaithyanathan, S.: Thumbs up?: sentiment classification using machine learning techniques. In: Proceedings of the ACL-02 Conference on Empirical Methods in Natural Language Processing, vol. 10, pp. 79–86. Association for Computational Linguistics (2002)
22. Prabowo, R., Thelwall, M.: Sentiment analysis: a combined approach. J. Inf. **3**(2), 143–157 (2009)
23. Salvetti, F., Reichenbach, C., Lewis, S.: Opinion polarity identification of movie reviews. In: Shanahan, J.G., Qu, Y., Wiebe, J. (eds.) Computing Attitude and Affect in Text: Theory and Applications, pp. 303–316. Springer, Netherlands (2006)
24. Schmeling, M.: Investor sentiment and stock returns: some international evidence. J. Empirical Finance **16**(3), 394–408 (2009)
25. Stone, P., Dunphy, D.C., Smith, M.S., Ogilvie, D.: The general inquirer: a computer approach to content analysis. J. Reg. Sci. **8**(1), 113–116 (1968)
26. Taboada, M., Anthony, C., Voll, K.: Methods for creating semantic orientation dictionaries. In: Proceedings of the 5th Conference on Language Resources and Evaluation (LREC 2006), pp. 427–432 (2006)
27. Tetlock, P.C.: Giving content to investor sentiment: the role of media in the stock market. J. Finance **62**(3), 1139–1168 (2007)
28. Tong Richard, M.: An operational system for detecting and tracking opinions in online discussions. In: Working Notes of the ACM SIGIR Workshop on Operational Text Classification, New York, pp. 1–6 (2001)
29. Turney, P.D.: Thumbs up or thumbs down?: semantic orientation applied to unsupervised classification of reviews. In: Proceedings of the 40th Annual Meeting on Association for Computational Linguistics, pp. 417–424. Association for Computational Linguistics (2002)

30. Turney, P.D., Littman, M.L.: Measuring praise and criticism: inference of semantic orientation from association. ACM Trans. Inf. Syst. (TOIS) **21**(4), 315–346 (2003)
31. Wilson, T., Wiebe, J., Hoffmann, P.: Recognizing contextual polarity in phrase-level sentiment analysis. In: Proceedings of the Conference on Human Language Technology and Empirical Methods in Natural Language Processing, pp. 347–354. Association for Computational Linguistics (2005)
32. Xia, R., Zong, C., Li, S.: Ensemble of feature sets and classification algorithms for sentiment classification. Inf. Sci. **181**(6), 1138–1152 (2011)
33. Zhou, L., Chaovalit, P.: Ontology-supported polarity mining. J. Am. Soc. Inf. Sci. Technol. **59**(1), 98–110 (2008)

On the Quality of Annotations
with Controlled Vocabularies

Heidelinde Hobel and Artem Revenko[✉]

Semantic Web Company, Vienna, Austria
{h.hobel,a.revenko}@semantic-web.at

Abstract. Corpus analysis and controlled vocabularies can benefit from each other in different ways. Usually, a controlled vocabulary is assumed to be in place and is used for improving the processing of a corpus. However, in practice the controlled vocabularies may be not available or domain experts may be not satisfied with their quality. In this work we investigate how one could measure how well a controlled vocabulary fits a corpus. For this purpose we find all the occurrences of the concepts from a controlled vocabulary (in form of a thesaurus) in each document of the corpus. After that we try to estimate the density of information in documents through the keywords and compare it with the number of concepts used for annotations. The introduced approach is tested with a financial thesaurus and corpora of financial news.

Keywords: Controlled vocabulary · Thesaurus · Corpus analysis · Keywords extraction · Annotation

1 Introduction

The interaction between corpora and controlled vocabularies has been investigated for a long time. With today's shift to semantic computing, the classical interplay between controlled vocabularies and corpora has intensified itself even further. The research goes in both directions: improving the processing of the corpus using controlled vocabularies (for example, query expansion [10] or word sense disambiguation based on thesauri [6]) and improving the controlled vocabularies using corpus analysis [1,2]. We focus on the second direction. As application areas are created faster than the general semantic economy can keep up with, targeted and automated improvements of thesauri are of utmost importance. On the one hand, many authors agree that so far one cannot rely on a completely automatic construction or even extension of controlled vocabularies [7,8]. On the other hand, the construction of a controlled vocabulary by an expert without the aid of the corpus analysis is also hardly possible. Therefore, the most reasonable scenario is to enable the expert to curate the construction process based on the results of the corpus analysis.

We investigate the scenario when thesauri are used as a source for concepts, and all the concepts found in the documents are used as annotations.

© Springer International Publishing AG 2016
A. Satsiou et al. (Eds.): IFIN and ISEM 2016, LNCS 10078, pp. 98–114, 2016.
DOI: 10.1007/978-3-319-50237-3_4

This paper is a preliminary report on our work in progress. This research is carried out in frames of the PROFIT[1] project.

Contribution. We introduce a measure of the quality of annotations. This measure is applicable to a pair of a thesaurus and a corpus; the value of this measure describes how well the two elements fit each others.

Structure of Paper. This paper is structured as follows. In Sect. 2 we introduce the reader to the topic of controlled vocabularies and their use for annotations. In Sect. 3, we describe our proposed approach for measuring the quality of annotations. We outline our performed case study in Sect. 4 and present our results in Sect. 4.4. We conclude the paper in Sect. 5 and present an outlook of future work in Sect. 6.

2   Controlled Vocabularies

Let a *label* L be a word or a phrase (a sequence of words). Here we understand *word* in a broad sense, i.e. it may be an acronym or even an arbitrary sequence of symbols of a chosen language. Let a *concept* C denote a set of labels. Usually we represent a concept by one of its labels that is chosen in advance (preferred label). A *controlled vocabulary* $\mathbb{V}$ is a set of concepts. We say that a *thesaurus* is a controlled vocabulary with additional binary relations [4] between concepts:

broader B a transitive relation, cycles are not allowed;
narrower N is an inverse of B;
related R an asymmetric relation.

Note that any thesaurus is a controlled vocabulary.

The concepts as defined here capture some features of a concept defined in SKOS [3]. We may consider SKOS thesauri as an instantiation of the thesauri defined here.

We understand *annotation* as a metadata attached to data; for example, a comment to a text or a description of an image. One may come up with different possible options for introducing annotations. For example, techniques for summarization of texts [11] or various techniques of text mining [15] offer multiple methods for introducing annotations. In this work we consider only the annotations that are done with a controlled vocabulary, i.e. only the concepts from a fixed controlled vocabulary can be used in annotations. Therefore, an annotation with a controlled vocabulary is a set of concepts from the controlled vocabulary. Among the advantages of making annotations with controlled vocabulary are the following:

---

[1] projectprofit.eu.

Consistency. Even if the annotations are done by different annotators (humans or algorithms) they are still comparable without any further assumptions or knowledge;

Translation. Though we cannot manipulate the annotated data, the annotations themselves may be translated, moreover, the annotator and the consumer may not even speak a common language;

Control over Focus. With the control over the controlled vocabulary one has the control over the focus of annotations, hence controlling the individual aspects of the data that require attention;

Alternative Labels. Thanks to multiple labels that may be introduced for a concept one may discover the same entity in different annotations even if the entity is represented by different synonyms.

Moreover, one may introduce additional knowledge about the concepts in the vocabulary in order to be able to perform advanced operations with annotations. For example, if one introduce a proximity measure over distinct concepts than one may compute finer similarity measure over annotations. Therefore, we may say that annotations with controlled vocabularies may be seen as a proxy for performing operations over data.

Though different types of data are important for the applications, the textual data offers unique chance to improve the thesaurus from the data. Moreover, in this case the annotations and the data are in the same format. The current state-of-the-art in combined approaches of thesauri and corpora allows to annotate texts in the corpora according to the described concepts in the thesauri and to employ phrase extraction from the corpora to identify new concepts for the thesauri.

3   Method

Goal. In this paper, we study an approach to assess and measure the quality of fit between a thesaurus and a corpus. The fit is understood in the sense of suitability of a given thesaurus to annotate a given corpus. The annotations are done automatically through finding the concepts from the thesaurus in the corpus.

We approach the goal via measuring the number of annotations found in each text. However, this number should correlate with the amount of information contained in the text. One way to assess the amount of information is to find keywords on a corpus level and check how many of those are contained in the text [13, 14].

3.1   Keywords

We utilize different methods to find keywords.

Mutual Information. Mutual information provides information about mutual dependence of variables and can be used to estimate if two or more consecutive words in a text should be considered a compound term that is formed of these words [9]. The idea is that if words are independent then they will occur together just by chance, but if they are ordered or occur together more often than expected then they are dependent. The definition of the mutual information score is as follows:

$$\mathrm{MI} = \log_2 \frac{P(t_{12})}{P(t_1)P(t_2)}, \; P(t_{12}) = \frac{f_{12}}{n_b}, \; P(t_i) = \frac{f_i}{n_s}, \tag{1}$$

where f_{12} is the total number of occurrences for the bigram t_{12}, n_b is the number of bigrams in the corpus, f_i is the total number of occurrences for the i-th word in the bigram, and n_s is the number of single words in the corpus. Analogously, the MI for n words can be calculated as follows:

$$\mathrm{MI} = \log_2 \frac{P(t_{1,\ldots,n})}{P(t_1)\ldots P(t_n)}. \tag{2}$$

Content Score. The idea of the content score is that terms that do not appear in most of the documents of a collection but when they appear in a document they appear a number of times are relevant to define the content of such documents. These terms are potentially relevant for thesaurus construction and one simple way to test to which degree a term falls in this category is to use a Poisson distribution. The idea is to predict based on this distribution the document frequency df based on the total frequency f of term i. If the predicted number is higher then the terms are accumulated in a lower number of documents than one would expect based on a random model. The difference between observed frequency and predicted one is what indicates (in an indirect way) the content character of a term. The probability that a term i appears in at least one document in a corpus is defined as follows:

$$p(\geq 1; \lambda_i) = 1 - e^{-\lambda_i}, \; \lambda_i = \frac{f_i}{n_d}, \tag{3}$$

where f_i is the total frequency of the term i in the whole corpus and n_d is the number of documents in the corpus. The inverse document frequency (IDF) of the term i is then defined as follows:

$$\mathrm{IDF}_i = \log_2(\frac{n_d}{df_i}), \tag{4}$$

where df_i is the document frequency of term i. The residual IDF (RIDF) for term i that is used to express the difference between predicted and observed document frequency:

$$\mathrm{RIDF}_i = \mathrm{IDF}_i - \log_2(p(\geq 1; \lambda_i)), \tag{5}$$

Final Score. As the final score the combination of the mutual information, the residual IDF, and the concept frequency is used. We filter the keywords according to this score and take different thresholds. As follows from the results represented in Sect. 4.4 the behavior is not dependent on the exact threshold. We use the obtained number of keywords for each text and the sum of the scores of the keywords as the measure of information contained in this text.

4   Case Study

In this section we describe the experimental setup, the used data, and the result of the experiment.

4.1   Data

We ran our experiment on

Thesaurus STW Economics[2];
Corpora The financial corpora extracted from investing.com.

Corpus. The chosen corpus for our analysis consists of 39, 157 articles obtained from the financial news website investing.com. The articles are divided into three categories:

eur-usd-news[3] 19, 119 articles;
crude-oil-news[4] 14, 204 articles;
eu-stoxx50-news[5] 5, 834 articles.

We collected the news articles by a customized parser that starts at the overview pages and dives into the subpages, storing the title as well as the full text. The corpora contains high quality financial articles and has been approved by financial experts as being representative corpora for the considered field. The news articles span from 2009 till 2016.

Thesaurus. STW Thesaurus for Economics [5,12] is a controlled, structured vocabulary for subject indexing and retrieval of economics literature. The vocabulary covers all economics-related subject areas and, on a broader level, the most important related subjects (e.g. social sciences). In total, STW contains about 6, 000 descriptors (key words) and about 20, 000 non-descriptors (search terms). All descriptors are bi-lingual, German and English (Table 1).

In frames of PROFIT project the STW thesaurus was extended. The figures for the extended thesaurus can be found in Table 2. This extension provides us

---

[2] http://zbw.eu/stw/.
[3] http://www.investing.com/currencies/eur-usd-news.
[4] http://www.investing.com/commodities/crude-oil-news.
[5] http://www.investing.com/indices/eu-stoxx50-news.

Table 1. STW thesaurus for economics statistics: original

Number of concepts	6521
Number of broader/narrower relations	15892
Number of related relations	11008

Table 2. STW thesaurus for economics statistics: extended

Number of concepts	6831
Number of broader/narrower relations	16174
Number of related relations	21008

with a unique opportunity to test the introduced measure. Since the extension was performed based on the corpus analysis and suggested keywords and was done by an expert in the domain we may suppose that the extension is sound and improves the fit between the corpus and the thesaurus.

4.2 Technology Stack

Our semantic technology stack includes term extraction, term scoring, and concept tagging. In the process of term extraction and concept tagging we lemmatize the tokens to improve the accuracy of the methods. We use PoolParty semantic suite[6] to perform all the mentioned tasks.

4.3 Workflow

The proposed workflow (see Fig. 1) consists of four main steps:

1. A web crawler automatically collects articles provided by financial news websites.
2. PoolParty is utilized to extract concepts, terms, and scores.
3. The numbers of found concepts and terms are obtained.

The results of this workflow are directly used to support decision making of experts.

4.4 Results

In this section, we present the results of the described experiment.

---

[6] https://www.poolparty.biz.

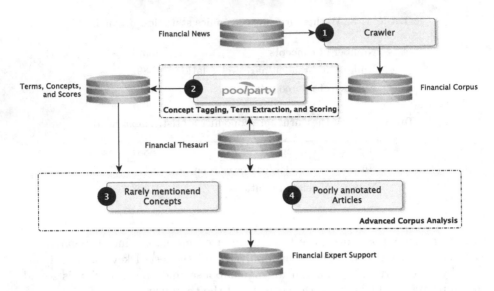

Fig. 1. Processing workflow

Counting Keywords. We start with comparing the number of keywords with scores above some threshold and a number of extracted concepts. In Figs. 2, 3, 4, 5. In the upper right corner of each figure the correlation coefficient R^2 and the slope of the line are presented. For all thresholds the overall picture remains the same, the correlation coefficient for the extended thesaurus is slightly larger than the same coefficient for the original thesaurus. The difference in the slope of the line is more significant.

Counting Sums of Scores of Keywords. Next we investigate the dependency between the sum of the scores of the keywords above certain threshold and the number of extracted concepts. The correlation coefficients for the extended thesaurus are larger; this indicates that the extended thesaurus is better suitable for the annotation of the given corpora.

It is worth noting that the number of extracted keywords and, hence, the sums of scores are significantly smaller for the extended thesaurus as some of the terms became new concepts and are not counted anymore (Figs. 6, 7, 8 and 9).

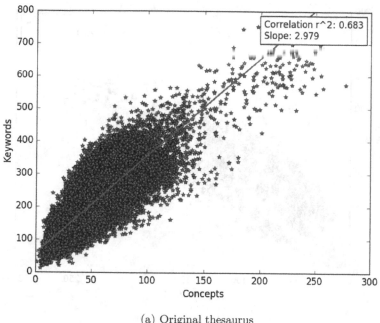

(a) Original thesaurus

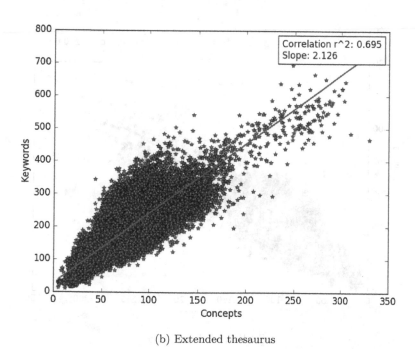

(b) Extended thesaurus

Fig. 2. Number of keywords vs number of concepts, threshold = 1

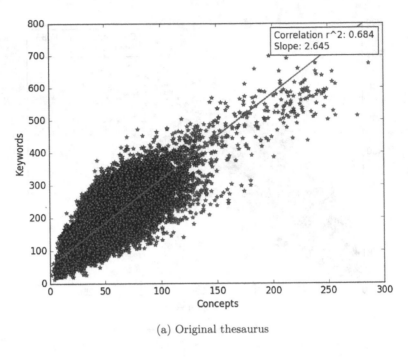

(a) Original thesaurus

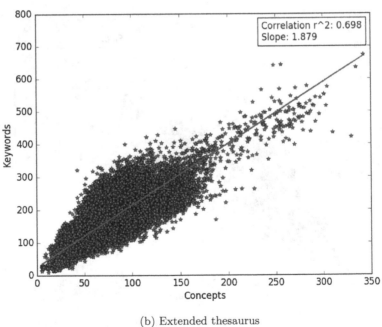

(b) Extended thesaurus

Fig. 3. Number of keywords vs number of concepts, threshold = 2

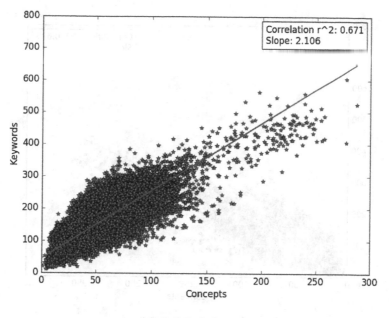

(a) Original thesaurus

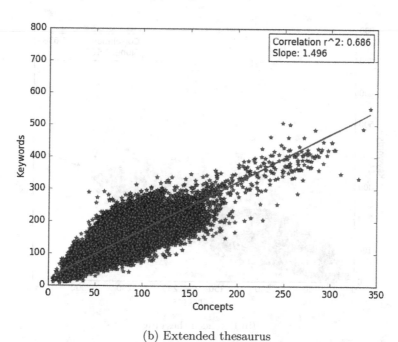

(b) Extended thesaurus

Fig. 4. Number of keywords vs number of concepts, threshold = 5

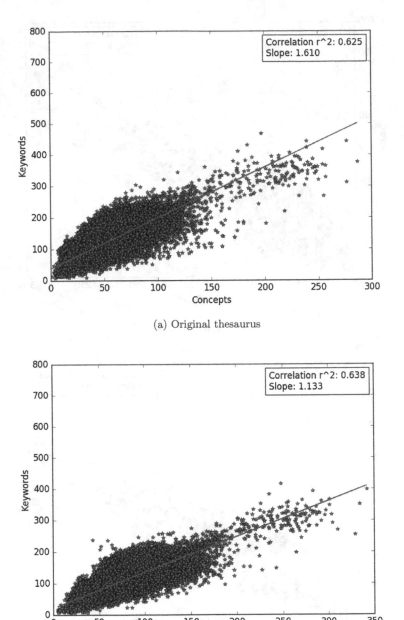

(a) Original thesaurus

(b) Extended thesaurus

Fig. 5. Number of keywords vs number of concepts, threshold = 10

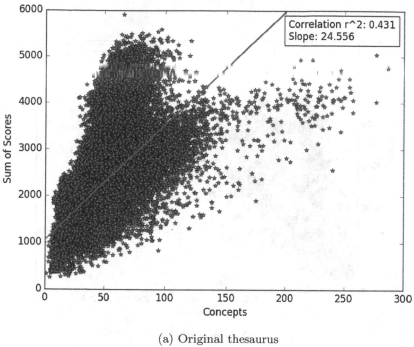

(a) Original thesaurus

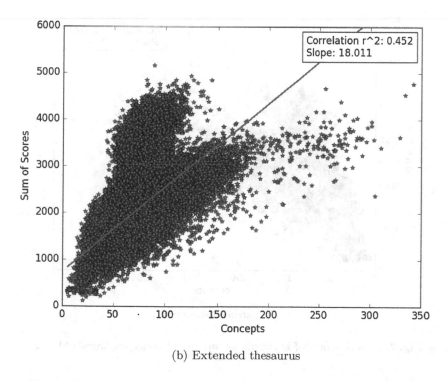

(b) Extended thesaurus

Fig. 6. Sum of scores of keywords vs number of concepts, threshold = 1

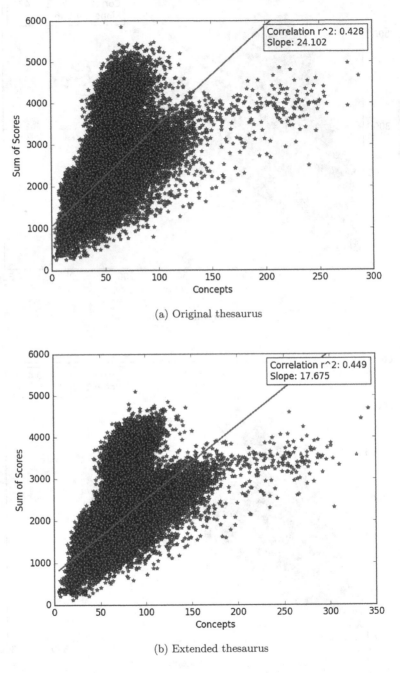

(a) Original thesaurus

(b) Extended thesaurus

Fig. 7. Sum of scores of keywords vs number of concepts, threshold = 2

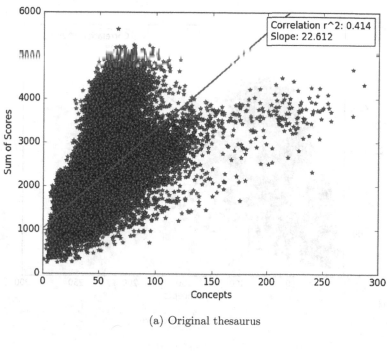

(a) Original thesaurus

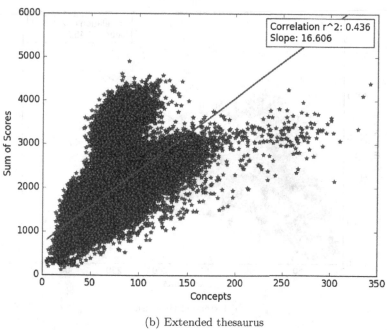

(b) Extended thesaurus

Fig. 8. Sum of scores of keywords vs number of concepts, threshold = 5

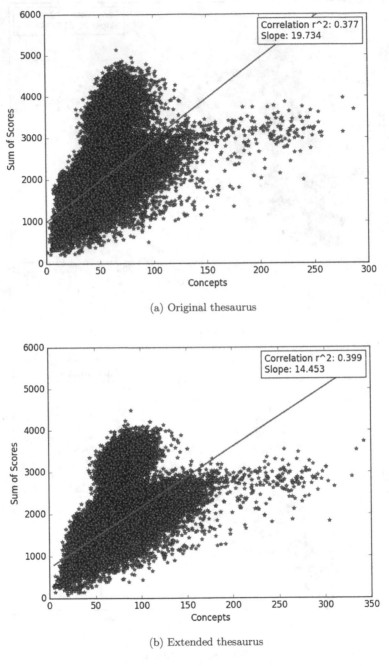

(a) Original thesaurus

(b) Extended thesaurus

Fig. 9. Sum of scores of keywords vs number of concepts, threshold = 10

5 Conclusion

Without automated support in the classical interplay between thesauri and corpora, a domain expert may use only his intuition in order to judge if a thesaurus is ready for being used in corpus analysis. We introduce a measure to assist the expert in deciding when to stop the development of thesaurus. The experimental evaluation shows promising results of the proposed approach to evaluate the fit of thesauri to the corpora. The presented approach is part of on-going work but we believe it provides a good foundation for future improvements.

6 Future Work

1. Investigate the plots of the number of extracted concepts vs the length of the document (in symbols and in words);
2. Investigate other measures of the keywords (for example, pure RIDF);
3. Investigate the plots for each topical corpus independently (eur/usd, oil prices, eustoxx);
4. Introduce not only quantitative measure of the annotation (number of concepts), but also a qualitative one (the relations between concepts);
5. Weight the concept occurrences using, e.g., IDF;
6. Investigate the possibility to find incorrect annotation due to, e.g., incorrect disambiguation;
7. Evaluate the results with further corpora and thesauri.

Acknowledgements. We would like to thank Ioannis Pragidis for his work on improving the thesaurus, pointing us to the relevant data, and sharing his deep expertize in the subject domain.

References

1. Ahmad, K., Tariq, M., Vrusias, B., Handy, C.: Corpus-based thesaurus construction for image retrieval in specialist domains. In: Sebastiani, F. (ed.) ECIR 2003. LNCS, vol. 2633, pp. 502–510. Springer, Heidelberg (2003). doi:10.1007/3-540-36618-0_36
2. Aussenac-Gilles, N., Biébow, B., Szulman, S.: Revisiting ontology design: a method based on corpus analysis. In: Dieng, R., Corby, O. (eds.) EKAW 2000. LNCS (LNAI), vol. 1937, pp. 172–188. Springer, Heidelberg (2000). doi:10.1007/3-540-39967-4_13
3. Bechhofer, S., Miles, A.: Skos simple knowledge organization system reference. In: W3C recommendation, W3C (2009)
4. Birkhoff, G.: Lattice Theory, 3rd edn. Am. Math. Soc., Providence (1967)
5. Borst, T., Neubert, J.: Case study: publishing stw thesaurus for economics as linked open data. In: W3C Semantic Web Use Cases and Case Studies (2009)
6. Jimeno-Yepes, A.J., Aronson, A.R.: Knowledge-based biomedical word sense disambiguation: comparison of approaches. BMC Bioinform. **11**(1), 1–12 (2010)
7. Kacfah Emani, C.: Automatic detection and semantic formalisation of business rules. In: Presutti, V., d'Amato, C., Gandon, F., d'Aquin, M., Staab, S., Tordai, A. (eds.) ESWC 2014. LNCS, vol. 8465, pp. 834–844. Springer, Heidelberg (2014). doi:10.1007/978-3-319-07443-6_57

8. Levy, F., Guisse, A., Nazarenko, A., Omrane, N., Szulman, S.: An environment for the joint management of written policies and business rules. In: 2010 22nd IEEE International Conference on Tools with Artificial Intelligence, vol. 2, pp. 142–149, October 2010

9. Magerman, D.M., Marcus, M.P.: Parsing a natural language using mutual information statistics. In: AAAI, vol. 90, pp. 984–989 (1990)

10. Mandala, R., Tokunaga, T., Tanaka, H.: Combining multiple evidence from different types of thesaurus for query expansion. In: Proceedings of the 22nd Annual International ACM SIGIR Conference on Research and Development in Information Retrieval, pp. 191–197. ACM (1999)

11. Mani, I., Maybury, M.T.: Advances in Automatic Text Summarization, vol. 293. MIT Press, Cambridge (1999)

12. Neubert, J.: Bringing the "thesaurus for economics" on to the web of linked data. In: LDOW, 25964 (2009)

13. Rose, S., Engel, D., Cramer, N., Cowley, W.: Automatic keyword extraction from individual documents. In: Berry, M.W., Kogan, J. (eds.) Text Mining, pp. 1–20. Wiley, New York (2010)

14. Shah, P.K., Perez-Iratxeta, C., Bork, P., Andrade, M.A.: Information extraction from full text scientific articles: where are the keywords? BMC Bioinform. 4(1), 1 (2003)

15. Tan, A.-H., et al.: Text mining: the state of the art and the challenges. In: Proceedings of the PAKDD 1999 Workshop on Knowledge Discovery from Advanced Databases, vol. 8, pp. 65–70 (1999)

ISEM 2016

Preface

It is our great pleasure to welcome you to the proceedings of the First International Workshop on Internet and Social media for Environmental Monitoring (ISEM) 2016 held in conjunction with the Third international conference on Internet Science (INSCI 2016), Florence, Italy.

The aim of ISEM 2016 was to present the most recent works in the area of environmental monitoring based on Web resources and user-generated content from social media (e.g., photos of the sky). The advancements in digital technologies and the high penetration of the Internet have facilitated the sharing of environmental information, such as meteorological measurements and observations of natural surroundings. Since the analysis of environmental information is critical both for human activities (e.g., agriculture, deforestation) and for the sustainability of the planet (e.g., nature conservation, green living, eco-driving, etc.), it is of great importance to develop techniques for the retrieval and aggregation of environmental information that is available over the Internet. Of particular interest is the exploitation of user-generated content, which, despite being of inconsistent quality in many cases, could contribute important information regarding areas that are not monitored by existing stations. In this context, ISEM 2016 focused on analysis, retrieval, and aggregation of environmental data from the Internet and user-generated content posted on social media, as well as on personalized services and decision support environmental applications (e.g., to suggest outdoor activities based on the current environmental conditions).

The Program Committee of ISEM 2016 accepted four papers covering a variety of topics structured in two sessions: (a) Web data analysis for environmental applications, and (b) air quality information based on user-generated content.

The first paper deals with compression of geodata for the development of real-time environmental applications. In particular, the paper presents a two-stage approach for the compression of Digital Elevation Model data and geographic entities for a mountain environment-monitoring mobile AR application.

The next research work conducted a social media analysis examining how social media outlets can focus information toward the desired demographics. Using a Facebook page about Citizens' Observatories, the authors reviewed indicators for evaluating public interest in social media content. The findings highlight the importance of up-to-date informational content, the use of visual content, and the role of features for interaction and dialogue to ensure user engagement with a Facebook page on environmental health.

The third paper deals with air quality estimation using collected multimodal environmental data. Specifically, it presents an open platform, which collects and fuses multimodal environmental data related to air quality from several sources including official open sources, social media, and citizens. The collection of such data aims at having complementary information, which is expected to result in increased geographical coverage and temporal granularity of air quality data.

Finally, the fourth paper deals with personalized air quality information based on open environmental data and user-generated information. The proposed application provides direct access to personalized and localized information on air quality (current, forecast, and historical), making use of diverse sources of large datasets of open air quality data, and crowdsourced information on the perception of app users about the current air quality.

In addition, Panagiota Syropoulou from DRAXIS delivered the keynote for the ISEM workshop by providing an overview of the hackAIR project, which aims at developing an open technology platform for citizens' observatories on air quality in order to raise collective awareness about the levels of human exposure to air pollution.

Finally, an open discussion took place focusing on public authorities' awareness of air quality and on participatory air quality monitoring. We would like to thank all the authors for submitting their work. We are also grateful to the Program Committee members for their effort and the high quality of the reviews, as well as to the members of the Organizing Committee of INSCI 2016 for constantly supporting us.

We hope that you find these proceedings a valuable reference and source of inspiration for future research on topics related to Web environmental monitoring.

September 2016   Stefanos Vrochidis
                                                       Symeon Papadopoulos
                                                      Christodoulos Keratidis
                                                       Panagiota Syropoulou
                                                             Hai-Ying Liu

Organization

General Chairs

Stefanos Vrochidis   Information Technologies Institute, Centre for Research and
                          Technology Hellas, Greece
Symeon Papadopoulos   Information Technologies Institute, Centre for Research and
                          Technology Hellas, Greece
Christodoulos Keratidis   DRAXIS Environmental S.A, Greece
Panagiota Syropoulou   DRAXIS Environmental S.A, Greece
Hai-Ying Liu   Norwegian Institute for Air Research (NILU), Norway

Program Committee

Berrin Yanikoglu	Sabanci University, Turkey
Pierre Bonnet	CIRAD, France
Hywel Williams	University of Exeter, UK
Anastasia Moumtzidou	CERTH-ITI, Greece
Kostas Karatzas	Aristotle University, Informatics Applications and Systems Group, Greece
Concetto Spampinato	University of Catania, Italy
Yiannis Kompatsiaris	CERTH – ITI, Greece
David Crandall	Indiana University, USA
Roman Fedorov	Politecnico di Milano, Italy
Ari Karppinen	Finnish Meteorological Institue, Finland
Yaela Golumbic	The Israeli Center of Research Excellence, Israel
Hervé Goëau	Cirad, France
Marina Riga	Aristotle University of Thessaloniki, Greece
Catherine Domingues	Institut Géographique National, France
Jaako Kukkonen	Finnish Meteorological Institute, Finland

Supported by

Horizon 2020 HackAIR project.

Compressing Web Geodata for Real-Time Environmental Applications

Claudio Cavallaro, Roman Fedorov^(✉), Carlo Bernaschina,
and Piero Fraternali

Dipartimento di Elettronica, Infomazione E Bioingegneria, Politecnico di Milano,
Piazza Leonardo da Vinci, 32, Milan, Italy
claudio.cavallaro@polimi.it,
{roman.fedorov,carlo.bernaschina,piero.fraternali}@mail.polimi.it

Abstract. The advent of connected mobile devices has caused an unprecedented availability of geo-referenced user-generated content, which can be exploited for environment monitoring. In particular, Augmented Reality (AR) mobile applications can be designed to enable citizens collect observations, by overlaying relevant meta-data on their current view. This class of applications rely on multiple meta-data, which must be properly compressed for transmission and real-time usage. This paper presents a two-stage approach for the compression of Digital Elevation Model (DEM) data and geographic entities for a mountain environment monitoring mobile AR application. The proposed method is generic and could be applied to other types of geographical data.

Keywords: Data compression · Augmented reality · Environment monitoring

1 Introduction

Outdoor augmented reality applications exploit the position and orientation sensors of mobile devices to estimate the location of the user and his/her device Field Of View (FOV) so as to overlay such view with information pertinent to the user's inferred interest and activity. These solutions hold the promise of engaging users to collect environmental data, exploiting the physical presence of the user in a position of interest and the possibility of sending his/her contextual, geo-referenced data to explain and support the data collection process.

A possible use case could be a mountain-specific Augmented Reality (AR) application [9], in which the user is steered towards collecting observations about specific mountains by overlaying his/her camera view with the identification, tracking distance, degree of interest, etc. of each mountain. The main challenges of outdoor mobile AR applications include providing an accurate estimation of user's information like current position, FOV and delivering the most relevant

This work was partially funded by the CHEST FP7 project of the European Commission.

A. Satsiou et al. (Eds.): IFIN and ISEM 2016, LNCS 10078, pp. 119–128, 2016.
DOI: 10.1007/978-3-319-50237-3_5

contextual meta-data, adapted to the changing view in real-time. Commercial applications, which operate mainly in the tourism field, simplify the problem by estimating the user's context based only on the device position and orientation sensors, irrespective of the content actually in view. Examples are sky maps, which show the names of sky objects based on the GPS position and compass signal. These approaches may provide information that does not match well what the user is seeing, due to errors in the position and orientation estimation or to the presence of objects partially occluding the view. These limitations prevent the possibility for the AR application to create *augmented content*. If the overlay of the meta-data onto the view is imprecise, it is not possible for the user to save a copy of the augmented view, e.g., in the form of an image with captions and meta-data associated to each object in view. Such augmented content could be useful for several purposes: archiving the augmented outdoor experience and the collected data, indexing visual content for supporting search and retrieval of the annotated visual objects, and even for the extraction of semantic information from the augmented content for environment analysis purposes. Furthermore, outdoor mobile AR applications must support field work in areas with low or absent network connectivity; this requires supporting offline usage, by letting the user download the potentially relevant meta-data before the data collection campaign. This requirement further advocates for intelligent data pre-processing to minimize the wait time, data storage, and data transmission time and cost, which could discourage citizens from using the application.

This paper describes the approach to real-time data management for the *SnowWatch* [2,8] outdoor mobile AR application, which supports mountain data collection campaigns, by contextualizing the user's view with the automatic overlay of geographical meta-data (peak name, altitude, distance from viewer, etc.). Unlike other systems (e.g., PeakFinder[1]), SnowWatch exploits a content-based reality augmentation algorithm [10], which takes as input not only the position and orientation of the user's device but also the content of the current view and meta-data about the region of interest. Such meta-data derive from a Digital Elevation Model (DEM), which is a 3D representation of the Earth's surface, stored at the server side. First, the DEM, the position and the orientation of the user are exploited to estimate a bi-dimensional projection of the panorama that should be in sight and to match it to the image currently captured by the camera. Second, meta-data about mountain peaks are transferred from the DEM to the camera view and superimposed in real time to it, so that the user can save an augmented image that integrates the contextual meta-data and the captured image. The augmentation process is executed in real-time and requires a policy for meta-data compression, transmission, decompression and caching, compatible with the capacity of the mobile device and able of ensuring the fluidity of the user's experience. The contributions of the paper can be summarized as follows:

- We introduce the problem of reality augmentation, specifically for mountain environment data collection purposes, and highlight the challenges of geographic meta-data management in a mobile AR context, in terms of latency,

---

[1] www.peakfinder.org.

data transfer cost, and support to disconnected usage, with a specific focus on DEM data.
- We describe multiple encoding and compression pipelines, which take advantage of the semantics of DEM data, for addressing such challenges. The best approach yields a compression ratio of 9.46 in average conditions and of 11.20 in favorable conditions (few peaks).
- We illustrate the application of the algorithms to a mobile AR application that supports real-time peak identification and augmented content generation.
- We report on the preliminary results of evaluating alternative data compression solutions in real outdoor experimental conditions.

The rest of the paper is organized as follows: Sect. 2 overviews previous work in the areas of data compression, outdoor AR mobile applications, and environmental monitoring applications; Section 3 states the problem of data management for outdoor AR application and introduces alternative algorithms addressing the challenges of this class of applications; Section 4 evaluates the algorithms for real-time DEM data management in different conditions; Section 5 concludes by discussing the generalization of the algorithms designed for DEM data to other environmental data management problems, presents the outcome of using annotated mountain images for the resolution of a real-world environmental problem, and provides an outlook about the next research objectives.

2   Related Work

Data compression is a prominent task in data science, for which many techniques have been developed over the years. Recently, the need of managing data in real time (e.g., for online applications) shifted the focus from the pure compression ratio optimization also to the minimization of the compression/decompression time. The leading data representation method for compression purposes is Huffman coding, which forms the basis of most subsequent approaches [20]. Compression techniques can be broadly classified in four major categories: derivatives of Lempel-Ziv-Welch [23], approaches based on statistical model prediction [4], on characters permutations [1], and on arithmetic coding [16]. In the case of DEM data, the seminal work [12] advocated the possibility of reducing data size with an initial data simplification stage, followed by compression with Huffman coding. Along this line, the work in [22] applied 3-points prediction, which yielded up to 80 % compression performance improvements; [14] applied an 8-points prediction, based on the Lagrange multipliers method, reaching 4 % compression performance improvements in comparison to [22]; [13] used a 24-points prediction algorithm in the pre-processing phase, achieving 40–60% improvement in comparison to gzip [6]. On the performance side, the work in [18] compares execution time of several lossless image compression algorithms, and applies them DEM data. Compression ratios and execution times for data in different geographic areas are collected and the optimal solution exploits a PNG compression method that supplements a deflate algorithm [5]. The work in [3] applies integer wavelet theory to pre-process data with both loss and loss-less methods, followed by arithmetic coding, which results in 80 % compression ratio, although with data loss.

In our work, we focus on the lossless compression of DEM data and follow the two-phase approach of previous methods, enhanced with heuristics in the data pre-processing phase for taking advantage of the knowledge about the data distribution; specifically, we exploit the terrain heterogeneity typical of DEM data in mountain ranges, to optimize both compression ratio and execution time in mountain regions, while preserving performance in other types of geographical areas.

3   Geodata Compression Algorithms

3.1   Problem Statement

The *SnowWatch* application addresses the problem of mobile AR for the enrichment of outdoor natural objects, specifically, mountain peaks. Restricting the focus to devices that support a bi-dimensional view, the mobile application receives as input a representation of the reality in which the user is embedded: a sequence of camera frames captured at a fixed rate, and the position and orientation of the device, acquired by the GPS and orientation sensors (e.g. magnetic sensor, gravity sensor, ...); the second input is the information about the possible objects present in a region of interest, as provided by the DEM of the geographical area surrounding the user; such information is used to render a 360° cylindrical virtual terrain view, which is aligned w.r.t. the current camera frame, to estimate the on-screen mountain peak positions. One of the biggest challenges is the requirement to operate in mountain regions, where internet connection may be unreliable. This calls for smart DEM pre-caching techniques, which can reduce consumption of both bandwidth and storage. The second challenge is the reduced computation power of mobile devices and the need to process DEM data in real time. The key to successfully address both problems is an efficient algorithm for DEM compression and decompression, guaranteeing a high level of compression and low decompression time, so that data can be transmitted faster, occupy less space in the device, and can be decoded quickly.

The DEM used in this work, provided by NASA, is the outcome of Shuttle Radar Topography Mission (SRTM) [7]; data can be seen as a regular grid overlaid on the Earth surface, where each grid point reports the height of the terrain in that position. We consider two different versions of SRTM DEM: SRTM1 at 1 arcsec resolution (which corresponds to ≈ 30 m in areas relatively far from the poles), and SRTM3 at 3 arcsec resolution (≈ 90 m). Both DEMs are provided as a series of tiles of $1'' \times 1''$, thus including $3601 \times 3601 / 1201 \times 1201$ points and having the size of 25, 327 KB and 2, 818 KB, respectively in case of SRTM1 and SRTM3. Data are saved as 16 bits in Little Endian encoding. Another intuitive way to see a DEM is a gray scale satellite image, where the color of every pixel is associated to the average height of the terrain in that area. Although several efficient lossy compression techniques can be used [11,17], DEM precision is vital for the correct outcome of peak identification (especially given other sources of error, such as device compass and GPS error). For this reason, we focus on lossless compression approaches, which allow us to compress a DEM and decompress it to its exact original state. Furthermore, to avoid selective downloading (e.g., ignoring data for areas of scarce mountaineering interest) and ensure full geographical coverage, the reduction should guarantee high compression in areas with both high and low altitude variance.

3.2 Pre-Compression Data Encoding

The goal of our work is to design an encoding technique applicable before compression (and symmetrically a decoding technique after decompression) [13], which exploits the specificity of DEM data and the purpose of the peak recognition application to improve compression ratio with minimal runtime overhead. SRTM data can be regarded as a 3D surface with maxima and minima that correspond to the mountain peaks and depressions. Although in optimal conditions (e.g., flat land) each point can be interpolated from its neighbors, in a realistic and common scenario, simple interpolation does not work well. A better method is to apply a predictive rule that estimates the value of a point from its neighbors [13]. Once estimated, the computed point value can be expressed as the difference between the original and the predicted value. Since difference values are smaller than absolute ones, the amount of information to compress is reduced. Although the idea is simple, it involves two non-trivial decisions: (1) which point to pick at start; (2) given a point, how many neighbors to use for predicting the value and how to select them. Differently from the described state-of-the-art methods, the proposed algorithm encodes DEM data recursively, as follows.

The procedure starts from the highest altitude point of the DEM tile to be encoded, and recursively calculates the value for the adjacent points in the possible eight directions. Each new value is calculated as the difference between the original point value and the average between the point treated in the preceding recursive step and another neighbor[2] (conventionally chosen as the next point in the current exploration direction). This transformation may produce both positive and negative values, which, due to two's complement encoding, are compressed inefficiency (e.g., the bit-to-bit difference between +1 and −1 is very high). Thus, a second step normalizes all values to positive, by adding to them the global minimum. This guarantees that elements close in terms of absolute value are also close in terms of bit encoding. Finally, the last step of the pre-processing exploits number encoding to create areas of the file with similar values. Each value in the file is a 16 bit (2 bytes) integer number. The high-byte of each cell is similar to its neighbors high-bytes, on the contrary the low-byte of each cell tend to be different from its neighbors low-bytes. Reorganizing the file putting all the first high-bytes first and all the low-bytes next we creates the above mentioned areas. Note that data decoding after decompression

---

[2] Trial-and-error experiments proved that increasing the number of neighbors for the average calculation decreases the total compression rate.

applies the same steps in reverse order (byte reorganization, de-normalization and recursive original value computation) and that the encoding is lossless.

This step allows to identify and remove resonance in the data, based on the terrain characteristic. The compression algorithm instead considers the data as a sequence of bytes and tries to exploit the remaining correlation of the data for a better compression. It is important to notice that the two techniques (proposed pre-compression step and the compression algorithm) are not interchangeable, because once the compression algorithm is applied, the data loose the characteristics needed to apply the pre-compression step.

3.3   Data Compression

The encoding described in the preceding section is generic and can be executed before any efficient compression algorithm; in this Section we describe the compression algorithms that produced the best results based on our experiments.

Currently, there are two classes of algorithms that perform optimally:

Lempel-Ziv compression methods (e.g., LZ77 [23]) work by substituting the occurrences of a specific string with a new value, composed by the distance from the first occurrence and the length of the string to copy. To further reduce the size of the new values, a dictionary table can be used, which associates values with shorter symbols. Such table can be stored as a header of the file and encoded. LZ77-based algorithms work well in presence of many repetitions, which are likely to occur after the described pre-processing phase. Furthermore, this family of algorithms provide a very fast decompression.

Prediction by Partial Matching (PPM) [4] algorithms consider the correlation between values (which, for example, could be linearly growing). They use N past values to predict the next one, trying to find the best relationship. During the prediction, an algorithm creates a ranking table in which more relevant repetitions are placed on top. Since our data values are already partially related (especially when the terrain is constant or linear) this method exploits its full potential. For several reasons PPM algorithms, however, tend to provide a worse performance, above all the decoding phase consuming a large amount of memory for past values and their frequency storage.

In our experiments, we compare two methods based on LZ77, namely, Deflate [5] and LZMA2 [15] and a PPM-based method called PPMd [21]. In addition, we also analyze Bzip2 [19], an algorithm based on Huffman coding [20], and the Burrows-Wheeler block sorting algorithm [1]. Bzip rearranges values inside a block to produce a more uniform sequence. Then the obtained sequences are substituted with symbols with Huffman coding.

4   Experimental Evaluation

4.1   Datasets

In compression problems, performance varies greatly with the nature of the data. For this reason, we experimented with different SRTM DEM examples, with variable characteristics:

- **ALPS1/ALPS3**: 34 square degree tiles (N43-N47, E5-E15) covering the whole Alpine region, extracted from SRTM1/SRTM3 DEM, for a total size of 98/881 MB respectively. Characterized by massive presence of mountains and high altitude variance.
  North America (NA). 48 square degree tiles (N52-N59, W97-W102) covering a part of Canadian Manitoba province, extracted from SRTM3 DEM, for a total size of 138 MB. Characterized by relatively low altitude variance.
- **ANDES**: 44 square degree tiles (S13-S16, W67-W77) covering the whole Andes region, extracted from SRTM3 DEM, for a total size of 127 MB. Characterized by mixed low and high altitude variance regions.

For each dataset we computed the compression ratio in two sessions: (1) using the four compression algorithms alone (PPMD, Bzip2, LZMA2 and DEFLATE); (2) using them after our pre-compression encoding step. We also measured the compression and decompression times for each test session. For the fairness of comparison, when the pre-compression encoding is used, its contribution is included in the total execution time. All experiments are performed on a Windows 10 desktop workstation, i7, 8 GB RAM.

Let C and C^* denote, respectively, the compression ratio (i.e., original size of data / compressed size) obtained by the algorithm alone and with pre-compression encoding; $\Delta C = (C^* - C)/C$ represents the improvement obtained by using the pre-compression encoding. We also define T^c and T^d as the compression and decompression time required by the algorithms. The star symbol ($*$) is applied also to execution times, to highlight the usage of pre-compression encoding; in this case, $\Delta T^c/\Delta T^d$ represent the improvement of using pre-compression encoding for the compression/decompression time, respectively.

Table 1 shows the obtained results on ALPS3 dataset, while Tables 2, 3 and 4 show the results respectively on ALPS1, NA and ANDES datasets.

Table 1. Experimental results for the ALPS3 dataset.

Algorithm	C	C*	ΔC (%)	Tc (s)	Tc* (s)	ΔTc (%)	Td (s)	Td* (s)	ΔTd (%)
PPMD	2.7	3.66	+35.56	23.16	13.90	+39.98	22.6	15.00	+33.62
Bzip2	2.96	3.29	+11.30	10.60	8.26	+22.07	4.00	3.30	+17.50
LZMA2	2.48	3.22	+29.83	23.20	11.50	**+50.43**	24.60	13.00	**+47.15**
DEFLATE	1.70	2.79	**+64.20**	7.30	7.00	+4.11	0.80	1.10	−37.50

The results show that the proposed pre-compression step provides a consistent improvement in compression efficiency in every scenario, varying from 6% to 106%, depending on the compression algorithm and the dataset. Although the best improvement occurs in regions with high mountain density and sensible altitude variance, the regions with low altitude variance still benefit from the pre-compression encoding. The execution time is generally improved too, with few exceptions (highlighted in red inside the result tables). However, it must be

Table 2. Experimental results for the ALPS1 dataset.

Algorithm	C	C*	ΔC (%)	Tc (s)	Tc* (s)	ΔTc (%)	Td (s)	Td* (s)	ΔTd (%)
PPMD	5.44	6.83	+25.55	96.5	88.1	+8.81	95.2	80.5	+15.96
Bzip2	5.12	5.65	+10.35	117.9	56,2	**+52.32**	29.8	22.5	+24.49
LZMA2	4.19	5.44	+29.91	161.1	102,3	+36.50	17.02	13.1	+23.03
DEFLATE	2.29	4.72	**+106.06**	43.0	74,5	-73.26	6.02	4.1	**+31.89**

Table 3. Experimental results for the NA dataset.

Algorithm	C	C*	ΔC (%)	Tc (s)	Tc* (s)	ΔTc (%)	Td (s)	Td* (s)	ΔTd (%)
PPMD	10.39	11.26	+8.42	3.7	4.8	-29.73	3.6	3.7	-2.78
Bzip2	9.22	9.84	+6.74	25.2	9.5	**+62.30**	6.46	4.2	**+34.98**
LZMA2	7.74	9.37	+21.04	35	21.5	+38.57	2.3	1.9	17.39
DEFLATE	5.57	8.39	**+50.62**	15	22	-46.67	0.8	0.7	+12.5

Table 4. Experimental results for the ANDES dataset.

Algorithm	C	C*	ΔC (%)	Tc (s)	Tc* (s)	ΔTc (%)	Td (s)	Td* (s)	ΔTd (%)
PPMD	2.71	3.57	+31.71	30.7	19.1	+37.78	35.6	21.2	**+40.50**
Bzip2	2.81	3.22	+14.56	12.34	10.01	+18.96	6.1	5.05	+17.20
LZMA2	2.98	3.20	+7.46	25.1	15.04	**+40.07**	5.0	3.4	+32.00
DEFLATE	1.79	2.89	**+61.65**	13.1	20.0	-53.84	1.1	1.0	+9.09

Table 5. Experimental results for the ALPS3 dataset on a mobile device.

Algorithm	Tc (s)	Tc* (s)	ΔTc (%)	Td (s)	Td* (s)	ΔTd (%)
PPMD	6.5	5.1	+21.53	10.36	4,63	**+55.31**
Bzip2	1.2	0.7	**+41.67**	0.48	0.35	+27.08
LZMA2	1.4	1.03	+28.57	1.12	0.71	+36.61
DEFLATE	1.1	0.9	+18.18	0.18	0.2	-11.11

Table 6. Experimental results for the NA dataset on a mobile device.

Algorithm	Tc (s)	Tc* (s)	ΔTc (%)	Td (s)	Td* (s)	ΔTd (%)
PPMD	2.33	2.2	+5.58	1.91	1.90	+0.5
Bzip2	1.5	0.7	**+53.33**	0.30	0.19	**+36.67**
LZMA2	1.9	1.2	+36.84	0.49	0.46	+6.1
DEFLATE	1.6	1.1	+31.25	0.06	0.08	-33.33

noted that the time increase occurs usually in cases where the pre-processing step achieves a high compression gain.

Since the goal of this work is to present a compression algorithm that could be efficiently executed also on mobile terminals, Tables 5 and 6 show the time performance measured on a mobile device (namely, Galaxy S5 with Snapdragon 801 @2.5 Ghz processor and 2 GB of Ram) on ALPS3 dataset (clearly, the compression performance remains invariant). As one may see, the benefits of using the proposed approach are even higher than on the desktop configuration. Although, due to the space constraints, we do not report the analogous performance on other datasets - they are similar to the ones presented in Tables 5 and 6.

5   Conclusions and Future Work

We presented an approach for the compression of DEM data and geographic entities for a mountain environment monitoring mobile AR application. We showed that a pre-compression encoding can be effectively applied, regardless of the underlying compression algorithm, and boosts the overall performance of the system both in terms of compression ratio and execution time. We tested the proposed approach with several state-of-art compression algorithms and DEM datasets with various characteristics. Although, as expected in compression problems, the final results vary sensibly w.r.t. the dataset and the compression algorithm, we obtained a constant improvement of the compression ratio, up to 50 %, and an average improvement of both compression and decompression time.

The simplicity and generality of the proposed approach allows one:

- To apply the approach with potentially any compression algorithm.
- To easily integrate the approach in an already existing compression pipelines.
- To process any kind of geographical data (besides the DEM), as long as they assign values denoting the property of interest to terrain models sampled on a regular grid (e.g., snow level, terrain temperature, vegetation health, etc.).

Our future work will address the usage of adaptive predictors instead of static ones, i.e. predictors that adapt flexibly to the terrain type and morphology.

References

1. Burrows, M., Wheeler, D.J.: A block-sorting lossless data compression algorithm (1994)
2. Castelletti, A., Fedorov, R., Fraternali, P., Giuliani, M.: Multimedia on the mountaintop: Using public snow images to improve water systems operation. In: Proceedings of the 24rd ACM International Conference on Multimedia. ACM (2016)
3. Chen, R., Li, X.: Dem compression based on integer wavelet transform. Geo-Spatial Inf. Sci. 10(2), 133–136 (2007)
4. Cleary, J.G., Witten, I.H.: Data compression using adaptive coding and partial string matching. IEEE Trans. Commun. 32(4), 396–402 (1984)
5. Deutsch, L.P.: Deflate compressed data format specification version 1.3 (1996)

6. Deutsch, L.P.: Gzip file format specification version 4.3 (1996)
7. Farr, T.G., Kobrick, M.: Shuttle radar topography mission produces a wealth of data. Eos Trans. Am. Geophys. Union **81**(48), 583–585 (2000)
8. Fedorov, R., Camerada, A., Fraternali, P., Tagliasacchi, M.: Estimating snow cover from publicly available images. IEEE Trans. Multimedia **18**(6), 1187–1200 (2016)
9. Fedorov, R., Frajberg, D., Fraternali, P.: A framework for outdoor mobile augmented reality and its application to mountain peak detection. In: Paolis, L.T., Mongelli, A. (eds.) AVR 2016. LNCS, vol. 9768, pp. 281–301. Springer, Heidelberg (2016). doi:10.1007/978-3-319-40621-3_21
10. Fedorov, R., Fraternali, P., Tagliasacchi, M.: Mountain peak identification in visual content based on coarse digital elevation models. In: Proceedings of the 3rd ACM International Workshop on Multimedia Analysis for Ecological Data, pp. 7–11. ACM (2014)
11. Franklin, W.R., Said, A.: Lossy compression of elevation data, pp. 29–41 (1996)
12. Kidner, D.B., Smith, D.H.: Compression of digital elevation models by huffman coding. Comput. Geosci. **18**(8), 1013–1034 (1992)
13. Kidner, D.B., Smith, D.H.: Advances in the data compression of digital elevation models. Comput. Geosci. **29**(8), 985–1002 (2003)
14. Lewis, M., Smith, D.: Optimal predictors for the data compression of digital elevation models using the method of lagrange multipliers
15. Pavlov, I.: 7z format. http://www.7-zip.org/7z.html
16. Rissanen, J., Langdon, G.G.: Arithmetic coding. IBM J. Res. Dev. **23**(2), 149–162 (1979)
17. Ruuvzika, J., Ruuvzika, K.: Impact of GDAL JPEG 2000 lossy compression to a digital elevation model, pp. 205–214 (2015)
18. Scarmana, G.: Lossless data compression of grid-based digital elevation models: a png image format evaluation. ISPRS Annals Photogrammetry Remote Sens. Spatial Inf. Sci. **2**(5), 313 (2014)
19. Seward, J.: bzip. 2 and libbzip. 2, version 1.0.5 a program and library for data compression (1996). http://www.bzip.org/1.0.5/bzip.2-manual-1.0.5.html#intro
20. Sharma, M.: Compression using huffman coding. IJCSNS Int. J. Comput. Sci. Netw. Secur. **10**(5), 133–141 (2010)
21. Shkarin, D.: PPMD-fast PPM compressor for textual data (2001)
22. Smith, D.H., Lewis, M.: Optimal predictors for compression of digital elevation models. Comput. Geosci. **20**(7), 1137–1141 (1994)
23. Ziv, J., Lempel, A.: A universal algorithm for sequential data compression. IEEE Trans. Inf. Theory **23**(3), 337–343 (1977)

Analysis of Public Interest in Environmental Health Information: Fine Tuning Content for Dissemination via Social Media

Hai-Ying Liu[1(✉)], Irene Eleta[2], Mike Kobernus[1], and Tom Cole-Hunter[2]

[1] Norwegian Institute for Air Research (NILU), Instituttveien 18, 2027 Kjeller, Norway
{Hai-Ying.Liu,Mike.J.Kobernus}@nilu.no
[2] Centre for Research in Environmental Epidemiology (CREAL), ISGlobal, Barcelona Biomedical Research Park, Dr. Aiguader 88, 08003 Barcelona, Spain
{Irene.Eleta,Tom.Cole.Hunter}@creal.cat

Abstract. This study conducts a social media analysis, defining a communication strategy for environmental health information, examining how social media outlets can focus information towards desired demographics. Using a Facebook page about Citizens' Observatories (COs), we reviewed indicators for evaluating public interest in social media content, and evaluated users' engagement with our COs page. The result is a practical method to promote and enhance the visibility of environmental health information. The major method is to exploit visual material to increase user engagement. The total sum of visits to the page was greatest when visual content was used. We found that environmental health content appeals to adults between 35–44 years of age, equally balanced between men and women. Our findings highlight the importance of up-to-date informational content, the use of visual content and the role of features for interaction and dialogue to ensure user engagement with a Facebook page on environmental health.

Keywords: Citizens' Observatories · Environmental health information · Public engagement · Social media · Scientific communication

1 Introduction

International environmental organizations, such as the United Nations Environment Programme (UNEP)[1], the Intergovernmental Panel on Climate Change (IPCC)[2], the European Environmental Agency (EEA)[3], and Friends of the Earth, are developing channels of scientific communication (e.g., News channels, various social media accounts, YouTube channels, etc.)[4] between environmental experts and the public,

---

[1] http://www.unep.org/NewsCentre/default.aspx?DocumentID=26827&ArticleID=35210.
[2] http://www.nature.com/nclimate/journal/v5/n4/full/nclimate2528.html?WT.ec_id=NCLIMATE-201504.
[3] www.eea.europa.eu/publications/eea-general-brochure/download.
[4] http://phys.org/news/2015-08-social-media-emergency.html.

© Springer International Publishing AG 2016
A. Satsiou et al. (Eds.): IFIN and ISEM 2016, LNCS 10078, pp. 129–146, 2016.
DOI: 10.1007/978-3-319-50237-3_6

exploring the use of social media platforms to foster public interest and promote engagement with their projects. They aim to develop communication channels that could help to empower communities and foster their active participation in environmental governance. The purpose of these communication channels is to generate public interest in environmental information and promote engagement with projects and organizations within the area of environmental health.

Social media is often used as an umbrella term for online platforms that enable the creation and exchange of user-generated content [1]. There are many social media platforms with different scopes and functionalities. Some are professional-oriented networking sites, such as LinkedIn, or are platforms for microblogging (e.g., Twitter), video sharing (e.g., YouTube), or knowledge sharing (e.g., Wikipedia) [2]. Today, more than 1.5 billion people around the world use social media to socialize, network, learn and share their interests [3], and this number of users is increasing [2]. The popularity of social media platforms and their inherent social networking structure enables the rapid diffusion of information [4]. Frequently, articles, videos or images are shared between thousands of people[5], generating information cascades known as "going viral" [4]. Social media can facilitate participation in public debates, civic engagement and organization of collective action [5].

A more generalist social networking site is Facebook. Currently, Facebook is the most popular social media service in the world, with 1.49 billion monthly active users as of June 30, 2015[6]. Companies and organizations are increasingly using Facebook pages and/or groups to communicate with their customers, members and target-groups [6]. This platform is already used for project dissemination, citizen recruitment and to help foster engagement in many environmental projects, e.g., WeSenseIt [7], CitCLOPS[7], and Clean Air in London[8].

This is the rationale of many programs using social media as online channels of communication between experts, citizens, organizations and authorities. For example, there are social media platforms for environmental professionals[9,10,11], such as Duke Environment[12]. In addition, lay people participate in environmental debates using social media or may share photos in real time on a variety of environmental issues, such as air pollution [8, 9], flood risk management [10] or forest fires [11]. Additionally, some organizations use social media to highlight their Corporate Social Responsibility (CSR) efforts, a form of branding or marketing, which includes environmental programs or campaigns [12, 13].

---

[5] https://www.networks.nhs.uk/nhs-networks/smart-guides/documents/Using%20social%20media%20to%20engage-%20listen%20and%20learn.pdf.

[6] http://newsroom.fb.com/company-info/.

[7] https://www.facebook.com/citclops/.

[8] http://cleanair.london.

[9] https://twitter.com/citisensemob.

[10] https://twitter.com/cobwebfp7?lang=en.

[11] https://www.facebook.com/WeSenseItProject.

[12] https://nicholas.duke.edu/programs/execed/courses/social-media-environmental-communications.

In recent years, there has been a boom in citizen science projects across the globe [14]. These projects allow members of the public to play an active part in monitoring and recording their environment. At the same time, Citizens' Observatories (COs) are an emerging approach for public participation in environmental monitoring, aimed at better observing, understanding, protecting and enhancing our environment [15]. COs appeared recently in the EU R&I (Research and Innovation) agenda, and involve the development of community-based environmental monitoring and information systems using innovative and novel Earth Observation (EO) applications in support of GEOSS (Global Earth Observation System of Systems). Five EU FP7 projects (CITI-SENSE[13], COBWEB[14], CitCLOPS[15], Omniscientis[16], and WeSenseIt[17] and one EMMIA project (Citi-Sense-MOB)[18] are currently developing novel technologies and applications in the domain of EO with the use of portable devices and enabling effective participation of citizens in environmental stewardship by actively involving them in community and policy priorities [15].

As part of the CITI-SENSE project, we created a COs Facebook page aimed at fostering communication between the COs-related project partners, stakeholders and users, to facilitate citizens' engagement, participation and network building, to disseminate information, and to improve participating projects' visibility.

In this paper, we proposed the use of social media analysis to inform the design of communication strategies in social media for citizens' engagement and participation in environmental health projects. The objectives of this study were to: (1) review existing indicators of public interest in social media content; (2) evaluate public interest in our pilot COs Facebook page and engagement with its content; and (3) provide recommendations for a social media communication strategy that is consistent with our aim.

In the following sections, we first provide a brief review of indicators of public interest in social media content, and then identify indicators and content that are related to the level of public engagement and empowerment in social media content, followed by a detailed description of our COs Facebook page. Furthermore, we present the metric analysis and results of public interest in environmental health information posted on our COs page. Finally, we discuss our results in the context of related literature and provide recommendations for increasing citizen engagement in environmental health projects.

2 Methods

2.1 Indicators of Public Interest in Social Media Content

Due to the rapid changes in the social media arena, our aim is not to present a list of current products since this would be soon outdated. Instead, we provide some examples

---

[13] https://www.citi-sense.eu.

[14] https://cobwebproject.eu/.

[15] https://www.citclops.eu.

[16] http://www.omniscientis.eu.

[17] https://www.wesenseit.eu.

[18] http://cwi.unik.no/wiki/CSM:Home.

of social media analysis tools to explain the information that can be obtained from them. Social media tools provide various indicators of engagement with social media content. Content can be textual posts, images, videos, files, and links to other websites/pages. The common indicators of public interest in social media content are: (i) page likes (number of people clicking 'like' for a page)[19]; (ii) post reach (number of people who viewed a post)[20,21]; (iii) post likes or favorites (number of people clicking 'like' or 'favorite' for a post; (iv) post follows (showing interest to receive updates for a particular post); (v) comments or replies (people leave comments for the post or reply other's comments for the post); (vi) sharing a post or retweeting (sharing or reposting the post with others).

2.2 Indicators of Public Engagement with Social Media Content

In addition to the indicators of public interest in social media content, social media analysis tools often provide a specified composed indicator, labelled "engagement," which typically includes post clicks, likes or favorites, comments or replies, and sharing instances. Some popular social media platforms, like Facebook, LinkedIn and Twitter, incorporate such indicators into their analysis tools. In addition, there are companies (e.g., Hootsuite[22], Buffer[23], SumALL[24], etc.) oriented to marketing, branding and campaign management that offer analysis of an organization's profile across many social media platforms (e.g., Social Media Today[25]). Depending on the personal data collected by a social media platform, there is also demographic information available on the visitors of social media pages as well.

2.3 Identifying and Sharing Content to Engage and Empower an Audience via Social Media

Citizens' engagement in environmental governance can take many forms. For that reason, engagement is often distinguished between high-level and low-level engagement of citizens. High-level involvement implies that citizens participate in environmental monitoring activities, contribute environmental data and help co-design project components. Low-level involvement implies that citizens consume information, for example, from social media platforms. Low-level involvement may be viewed as a prerequisite for enabling higher-level involvement, which demands good quality and inspiring content to trigger higher-level citizen participation.

---

[19] https://www.facebook.com/help/285625061456389.

[20] https://www.linkedin.com/help/linkedin/answer/26032?lang=en.

[21] https://support.twitter.com/articles/20171990#.

[22] https://hootsuite.com/.

[23] https://buffer.com/.

[24] https://sumall.com/.

[25] http://www.socialmediatoday.com/marketing/2015-03-10/9-best-free-social-media-analytics-tools.

For COs Facebook page (see next section), first, we shared relevant environmental health information with users in the hope of raising awareness (low-level engagement), which would later lead to higher-level engagement and then consequently, empowerment. In principle, better-informed citizens are empowered in the sense that they have the knowledge to take responsible actions locally. However, higher-level involvement, where citizens participate by contributing data and experiences, is essential for true empowerment. Empowered communities are those that participate in governance and the decision-making processes about issues that matter to them.

2.4   The Pilot Test: A Facebook Page for Citizens' Observatories

We started a pilot experiment creating a community profile for COs on Facebook. The COs Facebook page[26] was created as a non-partisan outlet for promoting any COs-related activity, whether initiated by COs projects[27], citizens, governmental bodies or related industry stakeholders. It is open to Small-to-Medium Enterprises (SMEs) that may want to promote new technologies, as well as projects or other initiatives that have COs-related component. The COs page constitutes a focal point for the promotion of COs worldwide and citizens are expected to contribute information, opinions, and to help build a community for influencing environmental decision-making.

We created the COs Facebook page on the 6 September 2013. Currently, it does not carry official branding for any particular project. Instead, we encourage mutual support for any COs related activity or project. In addition, the COs page serves as a dissemination tool for local COs-related initiatives that want to reach a wider audience. We update the COs page content regularly (i.e., minimum once or twice a week) and consider a SMART (Specific-Measurable-Attainable-Realistic-Time-Relevant-Time-bound) strategy when uploading content.

2.5   The Insights Tool

Facebook's Insights tool serves to evaluate the effectiveness and reach of information campaigns and to identify the type of posts and the post content that attracts more Facebook users. The analytical report produced by the Insights tool provides information on the total number of page visits, page likes (subscribers of the page), and posts' reach and engagement. There are different types of reach for Facebook posts: organic, viral and paid (e.g., Social Media Examiner[28]). The organic reach refers to the number of unique users who saw a post in their news-feed, or directly on the Facebook page. Viral reach counts all the people who saw a story about a page in their news-feed due to one or more of their friends liking, commenting or sharing the post on a page.

---

[26] https://www.facebook.com/int.cit.obs.

[27] http://citizen-obs.eu.

[28] http://www.socialmediaexaminer.com/facebook-reach-guide/.

Paid reach refers to the number of unique users who saw the post through an advertisement.

Additionally, the Insights tool provides demographic information of the users that engaged with the page or were reached by a page post. For example, there is data available on gender, age group, country, city, language, and time of the day when they connected to the Facebook page. In the Insights tool, data is presented in graphs and can be exported to Excel files[29].

2.6   Classification of Posted Content

With the objective of identifying which characteristics of the environmental content triggers more engagement and attention, we looked at the reach and engagement indicators provided by the Insights tool separately by type of media and topic (e.g., Air Pollution, Citizens' Observatories and Citizen Science). Based on contextual information, we also considered whether the posts refer to an event that has received attention from other media outlets.

First, we used the classification of posts in media types provided by the Insights tool: note (text only), photo with text, video with text, and link with text. Secondly, we selected the 10 posts with the highest reach and the 10 posts with the highest engagement values during the period 6 September 2013 to 19 October 2015 (from the day the COs page was created to the day COs page was analyzed). In total, we selected 15 posts. Five of them have high values both for post engagement and reach level. In Table 1, we synthesize the information about these 15 successful posts in six categories, namely: reach level, engagement level, post's type of media, content, topic of the content, and context.

3   Results

By October 2015, the COs Facebook page had been running for two years, with the total number of page likes reaching 274 (Fig. 1). Page likes grew gradually, and several COs-related SMEs (e.g., 1000001 Labs[30]) and projects (e.g., Citi-Sense-MOB) have started to use it for advertising and promoting their own environmental activities. Figure 2 shows the Insights tool overview of the COs page for a recent one-week period (From 13 October 2015 to 19 October 2015). This overview includes information on total page likes until 19 October 2015 (274, 0.7% increase in this recent week comparing from last week), new page likes within this week (2 new page likes, 66.7% decrease from last week), total post reach until 19 October 2015 (139, 4.8% decrease from last week), post reach within this week (119, 8.2% increase from last week), engagement within this week (12 users engaged, 33.3% decrease from the last week), and the five most recent posts with media type, reach level, engagement level and post name with published time.

---

[29] https://blog.kissmetrics.com/guide-to-facebook-insights/.
[30] https://www.facebook.com/1000001labs.

Table 1. Classification of the top ten posts with highest reach level, and top ten posts with highest engagement level for type of media, content, topics of the content, and context (5 posts that appear in the top ten fit into both criteria, so they overlap)

Post No	Total reach	Clicks	Likes, comments & shares	Type of media	Content	Topic	Context
1	364	23	10	Link with text	Link with text on 'mapping every city's most scenic routes, one photo at a time'	ICT / Citizen science / Smart city	A web portal named CityLab (http://www.citylab.com), which is dedicated to the people who are creating the cities of the future, and those who want to live there. Through sharp analysis, original reporting, and visual storytelling, focuses on the biggest ideas and most pressing issues facing the world's metro areas and neighborhoods
2	319	144	50	Video with text	NILU employee was interviewed by NRK about the Oslo Citizens' Observatories	Citizens' Observatories / Urban air pollution / Environmental awareness	NRK (Norwegian Broadcasting Corporation) news
3	309	16	8	Link with text	Link with text on 'inside Beijing's airpocalypse – a city made 'almost uninhabitable' by pollution'	Urban air pollution / Environmental awareness / Quality of life	This link takes users to a news shared in a well-known British national daily newspaper, The Guardian (http://www.theguardian.com), on the same day it was posted on the COs page
4	256	15	9	Link with text	Link with text on 'a Europe-wide project asks iPhone users to help monitor levels of pollution in major cities'	ICT / Citizen science / Citizens' Observatories / Urban air pollution	A well-known British national daily newspaper, The Guardian (http://www.theguardian.com)
5	232	23	10	Link with text	Link with text on 'wanted! an army of citizen scientists to tackle air pollution'	Citizen science / Citizens' Observatories / Urban air pollution / Environmental awareness	A well-known British national daily newspaper, The Guardian (http://www.theguardian.com)
6	188	19	11	Link with text	Link with text on 'with wearable devices that monitor air quality, scientist can crowdsource pollution map'	ICT / Citizen science / Citizens' Observatories / Urban air pollution / Crowdsourcing	A well-known online magazine named Smithsonian magazine which looking at the topics and subject matters researched, studied and exhibited (http://www.smithsonianmag.com)

(continued)

Table 1. (*continued*)

Post No.	Total reach	Clicks	Likes, comments & shares	Type of media	Content	Topic	Context
7	180	11	7	Note with link	An announcement about public air quality perception survey with web link from project CITI-SENSE	Citizen science Citizens' Observatories Urban air pollution Environmental awareness	A well-known EU project with its public air quality perception survey link (co. citi-sense.eu)
8	177	2	3	Note with link	Note with text on 'clear air, it is your move'	Citizen science Citizens' Observatories Air pollution Environmental health Quality of life	A slogan about European mobility week in year 2013 with a web link (http://www.aeidl.eu/en/news/latest-news/1024-european-mobility-week.html)
9	160	26	9	Video with text	Video with text about 'EC Citizens' Observatories projects'	Citizen science Citizens' Observatories Air pollution Water pollution Flooding Biodiversity Odor monitoring Environmental health Quality of life	A YouTube video
10	158	4	1	Link with text	A link with text about 'CITI-SENSE: do you want to know about the air you breathe?'	Citizen science Citizens' Observatories Urban air pollution ICT Environmental health Quality of life	A blog published on a well-known online blog about Global Health that is published on the Barcelona Institute for Global Health web page (http://www.isglobal.org)
11	89	33	3	Video with text	Video with text about 'citizen science and air quality monitoring in Tokyo'	Citizen science Urban air pollution	A global project blog (http://blog.safecast.org) which is working to empower people with data, primarily by mapping radiation levels and building a sensor network, enabling people to both contribute and freely use the data collected
12	124	31	2	Video with text	A video with text about 'TZOA wearable environ-tracker that monitors air pollution'	Citizen science Citizens' Observatories Urban air pollution	YouTube video (http://youtube.com/watch?v=lEv0Nw-eeVk)

(*continued*)

Table 1. (*continued*)

Post No.	Total reach	Clicks	Likes, comments & shares	Type of media	Content	Topic	Context
13	70	24	4	Video with text	A video with text about 'measure aerosols with your smartphone'	ICT / Citizen science / Citizens' Observatories / Air pollution	YouTube vide made by iSPEX project (https://www.youtube.com/watch?v=4_DO-5e4Y2w)
14	53	22	4	Video with text	A video with text about 'OpenIoT project which collects air quality data from mobile phone users that carry low-cost wearable sensors'	ICT / Citizen science / Citizens' Observatories / Urban air pollution	YouTube video made by OpenIoT project (https://www.youtube.com/watch?v=cKSEzqVgBXY)
15	117	23	2	Link with text	A link with a video and text from NRK about 'portable air monitor mounted on the electronic bike that monitor air pollution in Oslo'	ICT / Citizen science / Citizens' Observatories / Urban air pollution	NRK news

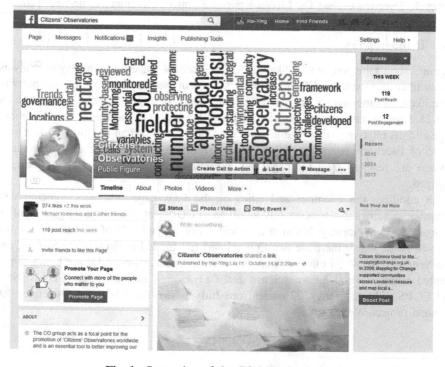

Fig. 1. Screenshot of the COs' Facebook page

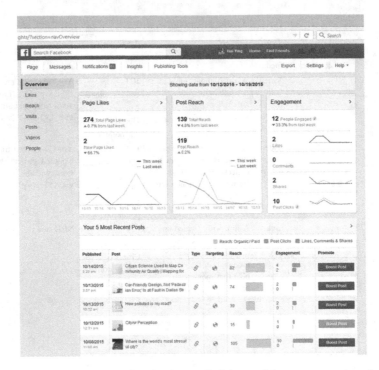

Fig. 2. Overview of the COs' Facebook page Insights tool for a recent one-week period

3.1   Demographics

Figure 3 shows the people who liked the COs Facebook page up to the day of analysis, the number of people the COs page posts served in the past 28 days, the people who have liked, commented on, or shared COs page posts or engaged with COs page in the past 28 days, by gender, age, country, city and language. In particular, COs Facebook page attracts relatively older age groups (i.e., 35–44 years) than the most frequent age group of Facebook users (i.e., 18–29 years). While a relatively higher percentage of women are reached by COs page and a relatively higher percentage of men are engaged.

In addition, the countries and cities of the COs page followers coincide with countries and cities that are currently involved in COs-related activities (e.g., Barcelona in Spain, Oslo in Norway, Milan in Italy, etc.). For example, 47 of our followers come from Norway, where there are two COs-related EU-funded projects, CITI-SENSE and Citi-Sense-MOB; 46 of our followers come from Spain, where COs are being implemented within several citizen science and COs-related projects (e.g., CITI-SENSE, CitCLOPS, iSPEX[31], etc.); 22 come from Italy where there is one COs-related EU founded project WeSenseIt.

---

[31] http://ispex-eu.org/.

Fig. 3. Screenshot of the age, gender and geographical demographics of people who liked, reached and engaged the COs' Facebook page, respectively

3.2 When are the COs Facebook Page Followers Online?

The Facebook Insights tool provides a weekly page update. We have checked all the weekly page updates and found that the COs page followers are online at predictable times as they follow a pattern. Figure 4 presents information about when (i.e., days and times) the COs page followers are online for a recent one-week period from 11 October

2015 to 17 October 2015. The time zone used for this is Central European Time (CET). We can see that there is not much variation among days (i.e., from Monday to Sunday), and on average about 90% of followers are online every day. Most COs followers are online from 9 am to 23 pm, with the least popular times for being online between 23 pm to 9 am, with a slump in followers' online presence between 21 pm to 9 am.

3.3   Interaction with Mass Media Outlets and Search Engines

The data points to a sudden increase in the total reach and engagement level after uploading relevant news from a TV broadcast on the COs page. There was a peak in the

Fig. 4.   Screenshot of when the COs' Facebook page followers are online

number of people (i.e., 319, 357% higher than the average post reach level) who read the post (total people reached by this post), 144 post clicks, 50 liked, commented and shared the posts (See Post No. 2 in Table 1, and Fig. 5). It occurred around the 10 February 2014, which was the date that the Norwegian public broadcast company (NRK) ran a news-story about the case study for the Oslo COs under both CITI-SENSE [15, 16] and Citi-Sense-MOB projects [15, 17]. From Fig. 5, we can see that there is a large gap for both total reach and engagement level before and after we uploaded the NRK news. For example, before the NRK news, on the 7 February 2014, we shared information about a citizen science project with a web link[32] and included the message: "be a citizen scientist for helping make mobility sustainable and our cities a little smarter". This post reached 62 people but only one engaged. Three days later, on the 10 February 2014, the NRK broadcast a news story about CITI-SENSE COs in Oslo, and we posted the NRK news on the COs Facebook page. This post reached 319 people,

---

[32] https://envirocar.org.

Fig. 5. Screenshot of the post reach and post engagement level of the COs' Facebook page during a predefined period

and 194 engaged. Later, on the 13 February 2014, we announced that our COs Facebook page had been promoted at an IoT-forum web site and included the web link[33]. This post only reached 52 people, and two were engaged (Fig. 5).

Figure 6 presents the number of times each of the COs page tabs (e.g., notes tab, info tab, events tabs and others) were viewed, and number of times people came to the page from another website (e.g., Google search engine, and other COs-related projects web pages) in the predefined time-period (i.e., from 3 February 2014 to 17 February 2014). For example, from Google search engine, the most common search terms that directed people to the COs Facebook page are "CITI-SENSE Citizens' Observatories", "Citizens' Observatories", and the combination of "citizen science" and "Citizens' Observatories".

---

[33] http://iotforum.org.

Fig. 6. Screenshot of the page and tab visits and external referrers of the COs' Facebook page

3.4 Media Types and Their Reach and Engagement Levels

To determine which type of media in a post is most popular for the users, we exported all 194 posts during the period from the 6 September 2013, to the 19 October 2015, including their type and reach and engagement level. Then, we calculated average post reach and engagement level for each media type (i.e., video with text, link with text, note, and photo with text) (Table 2). The results indicate that video with text have the highest average number of post reach and engagement, followed by photo with text, link with text, and note (See Table 2).

Table 2. Type of media, and the average post reach and engagement levels during the period 6 September 2013 to 19 October 2015

Post type of media	Post reach	Post clicks	Likes, comments and shares
Video with text	95	24	4
Photo with text	69	14	5
Link with text	65	5	2
Note (text only)	50	5	1

3.5   Most Popular Posts and Their Topics

There is an interaction effect between the post topics and the type of post that influence post popularity. That is, we are considering two factors that influence popularity, but we cannot isolate their actual effects. From Table 1, we can see that the most popular posts (both high reach and engagement levels) are those posts uploaded either as video with text from a TV broadcast (e.g., NRK), or as link with text from a famous online newspaper (e.g., The Guardian) with topics related to citizen science, Citizens' Observatories, air pollution, traffic-related air pollution. Nevertheless, within a total of 194 posts, the most popular topics (both high reach and engagement levels, See Table 1 on post Nos. 1, 2, 5, 6, 9) are Citizens' Observatories in the field of air pollution; citizen science for environmental awareness raising, especially for tackling air pollution issues; and ICT (Information and Communication Technologies) with smartphone and low-cost air monitor for mapping cities' air quality.

4   Discussion

The COs Facebook page is becoming a popular and intuitive platform for COs-related projects' dissemination, citizens' recruitment and citizens' empowerment. Moreover, as popularity grows people start to use the COs page as a resource for promoting their project or business. For example, SME 1000001 Labs uses the COs page to advertise that it is specialized in decision support, recommendation, personalization and IoT applied to the marine-environment and healthcare. Until 19 October 2015, the total page likes reached 274, and we anticipate a wider audience will start to rely on it for learning about what a Citizens' Observatory is, and what it can do for them. It should be mentioned that all 274 page likes are from organic reach, rather than paid reach through an advertisement. For the COs page, paid reach may help to increase the audience, who may help spread the message of the COs page. However, this paid reach may not really attract an audience interested in the COs page content *per se*. Based upon this consideration, the paid reach is irrelevant. This means attention must be directed at understanding what kind of content increases organic reach. Advice for increasing the organic and viral reach of Facebook Page posts typically point to the importance of varying the content and including visual and media-rich elements, creating shareable quote posts, and providing specific calls to action. Such recommendations point to how providing interesting content and engaging users are interlinked: content is a means to spark interest in citizens' participation.

Regarding the results of the COs Facebook page that indicated most Facebook users are on Facebook during the day, but before midnight (i.e., from 9 am to 23 pm), particularly between 13–15 pm and 19–22 pm, which are the most popular time for COs page, we suggest that social media (such as Facebook) page updates should be made during this particular time-period when the majority of page-followers are online and active. However, during these hours, it is expected that there is also more competition from other pages for reaching people, and it is hence worth experimenting with different scheduling approaches, for example, by using programs (e.g., Hootsuite, Buffer and SumALL) which allow users to schedule the posting of content to see what works more efficiently (e.g., scheduling the same posts at different times of the same day), and to explore which times increase the reach of posts and investing effort into finding interesting (relevant) content.

The observation that the COs Facebook page attracts a relatively older age group (e.g., 35–45 years) than the general Facebook user demographic may be linked to the topics of the content of COs that we have promoted, which may be of more interest to professionals and/or parents, rather than, for example, students. If relatively younger age groups and/or elderly are to be targeted, posts should be tailored to fit their content-preferences, in terms of both topic and the posts type of media. This requires further investigation.

Country and demographic information of the COs' page users may provide a lot of information about the success of the various COs' activities in specific countries/cities, targeting different gender and age groups. However, to be able to evaluate this, there needs to be SMART-targets[34] based on this kind of information, e.g., a target on the total number of followers, a target on involving followers from different age groups, cities and countries, and so on.

From approximately two years of managing the COs Facebook page, we have observed that page posts with visual or video content tend to generate more engagement than textual posts. We also observed that posts on specific topics, such as air pollution related COs or citizen science programs and activities, tend to reach more followers than COs in a broad sense. For this reason, we recommend to include a visual or graphical element in the posts, and post more on specific topics than broad concepts.

Facebook's Insights tool can help social media managers make sense of social media data and determine the best type of content to drive attention to their page and target specific demographics. However, the Insights tool has its drawbacks. For example, it provides weekly post updates, but does not provide a record of the total number of posts and only displays a limited number of posts with engagement information in a comparison table. These data are not available in tabular form to export to a Microsoft Excel spreadsheet, but can be calculated manually by the Facebook page manager. It does not detect the popular post topics and sentiment analysis of post (both positive and negative comments) (Stieglitz & Dang-Xuan [2]). Therefore, further development is needed to overcome its drawbacks and better facilitate the post analysis (comparison) by providing a tabular form of record of the total number of posts with engagement information. Furthermore, the increasing sophistication of social media analysis, like detecting influential users on social networks or tracking the posts of specific users (Stieglitz & Dang-Xuan [2]), calls for the inclusion of ethical protocols to protect users' privacy when designing the methods for analysis. For example, the analysis tools of Facebook, Twitter and LinkedIn provide aggregated results and it is not possible to single-out any individual.

One positive aspect of our current communication strategy through the COs Facebook page is that we gather attention to COs-related projects across national borders. On the other hand, previous studies have shown the importance of adapting environmental information to small geographic areas, tailored for locals, to foster

---

[34] http://www.i-scoop.eu/smart-social-media-listening-monitoring-s-m-r-t/.

citizens' engagement [18]. This attention between local and internationally oriented information requires a lot of research in itself.

The COs Facebook page represents a step forward with respect to the usual social media communication strategies of public health departments. Among those that have innovated by using social media to disseminate information to the public, many have not yet leveraged the interactive features that could engage a wider audience and increase the reach of their messages [19].

5 Conclusions

We created the COs Facebook page to engage with citizens, to facilitate networking with peers, promote the concept of COs, and disseminate COs-related activities and products for environmental monitoring. The Facebook Insights tool helped COs page managers to gain an understanding of public interest and social participation in this page, with the objective of making more informed decisions about the publication strategies of environmental health information.

In the COs Facebook page, via Insights' quantifiable metrics and data visualizations, we can conclude that most of the COs' Facebook followers are (i) from those cities and countries that are involved in COs-related activities, (ii) both women and men equally, and (iii) middle-aged adults. Interestingly, the COs Facebook posts with visual content tend to generate greater engagement than textual posts. Furthermore, we found that posting COs-related content that has been broadcast via TV increases engagement, and is one way to leverage the interaction factor between mass media and social media.

Apart from the type/form of information (i.e., text, video, link, etc.) that is being posted to the user, we believe that the actual content of the information is of great importance for the total number of users that visit, reach, or are interested for a post. Personalized information according to user needs can also have great impact in the achieved engagement and potential increase of public interest.

Acknowledgments. This work has been partly supported by CITI-SENSE (Development of sensor-based Citizens' Observatory Community for improving quality of life in cities), a Collaborative Project co-funded by the EU FP7-ENV-2012 under grant agreement no 308524.

References

1. Kaplan, A.M., Haenlein, M.: Users of the world, unite! The challenges and opportunities of social media. Bus. Horiz. **53**, 59–68 (2010)
2. Stieglitz, S., Dang-Xuan, L.: Social media and political communication: a social media analytics framework. Soc. Netw. Anal. Min. **3**, 1277–1291 (2013)
3. Das, B., Sahoo, J.S.: Social networking sites – a critical analysis of its impact on personal and social life. IJBSS. **2**, 222–228 (2014)
4. Sun, E., Rosenn, I., Marlow, C., Lento, T.: Gesundheit! Modeling Contagion through Facebook News Feed. Association for the Advancement of Artificial Intelligence (2009)

5. Segerberg, A., Bennett, W.L.: Social media and the organization of collective action: using twitter to explore the ecologies of two climate change protests. Commun. Rev. **14**, 197–215 (2011)
6. Rheingold, H.: Using participatory media and public voice to encourage civic engagement. In: Lance Bennett, W. (eds.): Civic Life Online: Learning How Digital Media Can Engage Youth. The John D. and Catherine T. MacArthur Foundation Series on Digital Media and Learning, pp. 97–118. The MIT Press, Cambridge (2008)
7. Lanfranchi, V., Wrigley, S.N., Ireson, N., Ciravegna, F., Wehn, U.: Citizens' observatories for situation awareness in flooding. In: Proceedings of the 11th International ISCRAM Conference, ISCRAM, pp. 145–154. (2014)
8. Kay, S., Zhao, B., Sui, D.: Can social media clear the air? A case study of the air pollution problem in Chinese cities. Prof. Geogr. **67**, 351–363 (2015)
9. Wang, S., Paul, M.J., Dredze, M.: Social media as a sensor of air quality and public response in China. J. Med. Internet Res. **17**, e22 (2015)
10. Wehn, U., Evers, J.: The social innovation potential of ICT-enabled citizen observatories to increase eParticipation in local flood risk management. Technol. Soc. **42**, 187–198 (2015)
11. Sutton, J., Palen, L., Shklovski, I.: Backchannels on the front lines: emergent uses of social media in the 2007 Southern California Wildfires. In: ISCRAM, pp. 624–632 (2008)
12. Curley, C.B., Noormohamed, N.A.: Social media marketing effects on corporate social responsibility. JBER **12**, 61–66 (2014)
13. Sandoval, M.: From Corporate to Social Media – Critical Perspectives on Corporate Social Responsibility in Media and Communication Industries. Routledge, New York (2014)
14. Catlin-Groves, C.L.: The citizen science landscape: from volunteers to citizen sensors and beyond. Int. J. Zool. **2012**, 1–15 (2012)
15. Liu, H.-Y., Kobernus, M., Broday, D., Bartonova, A.: A conceptual approach to a citizens' observatory - supporting community-based environmental governance. Environ. Health. **13**, 107 (2014)
16. Engelken-Jorge, M., Moreno, J, Keune, H., Verheyden, W., Bartonova, A., CITI-SENSE Consortium: Developing citizens' observatories for environmental monitoring and citizen empowerment: challenges and future scenarios. In: Proceedings of the Conference for E-Democracy and Open Government, pp. 49–60. CeDEM (2014)
17. Castell, N., Kobernus, M., Liu, H.-Y., Schneider, P., Lahoz, W., Berre, A.J., Noll, J.: Mobile technologies and services for environmental monitoring: the citi-sense-MOB approach. Clim. Change **14**, 370–382 (2015)
18. Adams, P.C., Gynnild, A.: Environmental messages in online media: the role of place. Environ. Commun. **7**, 113–130 (2013)
19. Thackeray, R., Neiger, B.L., Smith, A.K., Van Wagenen, S.B.: Adoption and use of social media among public health departments. BMC. Pub. Health. **12**, 242 (2012)

Towards Air Quality Estimation Using Collected Multimodal Environmental Data

Anastasia Moumtzidou[1], Symeon Papadopoulos[1], Stefanos Vrochidis[1],
Ioannis Kompatsiaris[1], Konstantinos Kourtidis[2], George Hloupis[3],
Ilias Stavrakas[3], Konstantina Papachristopoulou[4],
and Christodoulos Keratidis[4]

[1] Centre for Research and Technology Hellas - Information Technologies Institute,
Thessaloniki, Greece
{moumtzid,papadop,stefanos,ikom}@iti.gr
[2] Democritus University of Thrace, Xanthi, Greece
kourtidi@env.duth.gr
[3] Technological Education Institute of Athens, Athens, Greece
hloupis@teiath.gr, ilias@ee.teiath.gr
[4] DRAXIS Environmental Technologies Company, Thessaloniki, Greece
{k.papachristopoulou,keratidis.ch}@draxis.gr

Abstract. This paper presents an open platform, which collects multimodal environmental data related to air quality from several sources including official open sources, social media and citizens. Collecting and fusing different sources of air quality data into a unified air quality indicator is a highly challenging problem, leveraging recent advances in image analysis, open hardware, machine learning and data fusion. The collection of data from multiple sources aims at having complementary information, which is expected to result in increased geographical coverage and temporal granularity of air quality data. This diversity of sources constitutes also the main novelty of the platform presented compared with the existing applications.

Keywords: Environmental data · Air quality · Multimodal · Collection

1 Introduction

Environmental data is very important for human life and the environment. Especially, the environmental conditions related to air quality are strongly related to health issues (e.g. asthma) and to everyday life activities. Such data is measured by dedicated stations established by environmental organizations, which are usually made available through websites and services. Furthermore, the availability of low cost hardware sensors allows for the establishment of personal environmental stations by citizens. In parallel, the increasing popularity of social media has resulted in massive volumes of publicly available, user-generated multimodal content that can often be valuable as a sensor of real-world events [2]. This fact coupled with the rise of citizens' interest in environmental issues, has triggered

© Springer International Publishing AG 2016
A. Satsiou et al. (Eds.): IFIN and ISEM 2016, LNCS 10078, pp. 147–156, 2016.
DOI: 10.1007/978-3-319-50237-3_7

the development of applications that make use of social data for collecting environmental information and creating awareness about environmental issues.

To this end, this paper presents a new platform developed by the hackAIR project[1], for gathering and fusing environmental data and specifically Particulate Matter (PM) measurements from official open sources and user generated content (including social media communities) covering mostly urban areas but also rural areas provided that there is data for them. This platform aims to contribute towards individual and collective awareness about air quality and to stimulate sustainable behaviour with respect to it.

2 Relevant Initiatives

There are several initiatives including projects and applications that attempt to provide citizens with environment-oriented information collected from different data sources. Table 1 contains a detailed list of such initiatives. For the sake of space, we shall briefly mention only a limited number of them: (a) iSCAPE [20] that encapsulates the concept of smart cities by promoting the use of low cost sensors and the use of alternative solution processes to environmental problems, (b) the Amsterdam Smart Citizens Lab [7] that uses smartphones, smart watches, and wristbands, as well as open data and DIY sensors for collecting environmental data, (c) AirTick [7], which estimates air quality in Singapore by analysing large numbers of photos posted in the area, (d) PESCaDO [27] that focused on open environmental sources and provided users with personalized information, (e) PASODOBLE [26] that promotes air quality services integrating observations development of basic localised information for Europe, (f) CITI-SENSE [13], which provides air quality information based on portable and stable sensors, (g) CAPTOR [11], which aims at engaging a network of local communities for monitoring tropospheric ozone pollution using low-cost sensors, and (h) idokep [19] that provides weather information based on user generated data.

As far as the applications are concerned, the most interesting are: (a) Ubreathe [32] that provides current and forecast air quality as well as health advice for UK, (b) World Air quality [34] that reports Air Quality Index for 500 cities around the world using air quality monitoring stations, (c) AirForU [5] that provides Air Quality Index, hourly updates, one day forecast, historical exposure, and personalised tips, (d) Air Visual [3] that presents historical, real-time and forecast air quality data, including PM10, SO2, temperature using indoor and outdoor sensors, (e) AirCasting [4], which is an open-source platform that consists of wearable sensors that detect changes in your environment and physiology, including a palm-sized air quality monitor, an Android app, and wearable LED accessories, (f) Banshirne [10], that provides air pollution forecast based only from weather data, (g) Clean Air Nation [14] that provides air quality recommendations for special groups based on measurements coming from monitoring stations, (h) App for Kids to ID Asthma Attack Triggers [8], which

---

[1] www.hackair.eu.

collects data from smartphones and smart watches that identify risk factors and environmental triggers for asthma attacks, and (i) ActAQ! [1] that uses open data to provide air quality personalised information.

Compared to the aforementioned initiatives, the proposed platform combines data from various sources in an effort to benefit from the reliability of open official data, the abundance and high coverage of publicly available images posted through social media, the quality and consistency of images captured by users of the platform-oriented mobile app and the reliability of the measurements of low-cost open sensor devices for relatively large numbers of community users.

Table 1. Relevant initiatives

Type	Name of initiative
Project	iSCAPE [20]
	Amsterdam Smart Citizens Lab [7]
	AirTick [6]
	PESCaDO [27]
	PASODOBLE [26]
	CITI-SENSE [13]
	CAPTOR [11]
	idokep [19]
Application	Ubreathe [32]
	World Air quality [34]
	AirForU [5]
	Air Visual [3]
	AirCasting [4]
	Banshirne [10]
	Clean Air Nation [14]
	App for Kids to ID Asthma Attack Triggers [8]
	ActAQ! [1]

Apart from the aforementioned projects and applications, a work that is related to the proposed system is that of Tang et al. [31] that uses the EventShop software, which provides a generic infrastructure for analyzing heterogeneous spatio-temporal data streams regarding real-world events. However, EventShop focuses mainly on the fusion, and analysis of measurements coming from different sources and is not concerned with the retrieval and analysis of multimodal data.

3   System Architecture

The proposed system will collect PM measurements from various sources that will be processed according to their type (i.e. text, image). The sources that are foreseen are: (a) web-based official sources (contain data coming usually from national measurement networks, or local measurements networks by city authorities), (b) image-based sources, and (c) hardware-based sources. The aim of having different sources is to address the need for both reliable measurements (official sources satisfy this criterion) as well as large in number and broad coverage area measurements (user generated measurements expect to compensate for the lack of official data, e.g. areas not reached by the official sources such as rural areas).

As far as the web-based official sources are concerned, these involve publicly available open data found in environmental websites and services. Regarding image-based sources, they include publicly available geotagged images posted through platforms such as Instagram or Flickr, images captured by the users of the hackAIR mobile app and webcams. Finally, hardware-based sources involve low-cost open sensor devices assembled by citizens to monitor PM concentration.

The diversity of sources results in multimodal in nature input data that include images, unstructured and structured text and numeric values. Depending on the type of data, different analysis procedures are foreseen. Specifically, images (coming from social media, the mobile app, and webcams) will be processed using image analysis techniques and by measuring color ratios in the parts of the image containing sky an image-based air quality estimation in the form of an air quality index (e.g. low, high) will be produced. In the case of websites, the data is provided in unstructured format and thus information extraction techniques are required to extract the target information. Moreover, in the case of web services, the data is provided in structured format and thus no complex text processing is required. Finally, in the case of user-developed sensors, the sources provide numeric values. Eventually, all collected data is stored into a Sensor Observation Service (SOS) server repository [24]. Figure 1 depicts an overview of the hackAIR architecture.

In the remaining of the paper, we present the data sources used and the techniques that will be applied for retrieving data from these sources and the post-processing techniques.

4   Data Sources and Retrieval Techniques

4.1   Web-Based Official Sources

These include web services and websites that contain environmental measurements. The discovery and indexing of web services and webcams is conducted with the help of air quality experts, while the websites are discovered using domain specific web search and crawling, which can be divided into two main categories: the first is based on using existing general search engines to access the web and retrieve a first set of results, which are subsequently filtered with the aid of post-processing techniques [12,25]; the second is based on using a set of predefined websites and expanding them using focused crawling and with the help of machine learning techniques [37].

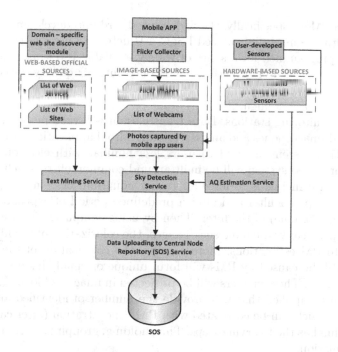

Fig. 1. System architecture. (Color figure online)

4.2 Image-Based Data Sources

Those include images found in: (1) media sharing platforms such as Flickr, (2) images captured from the mobile app and (3) webcams.

Regarding the images found in media sharing platforms such as Flickr, they offer the advantage of abundance and high geographic coverage. Initially, only geotagged images will be collected and thus for a known rectangle enclosing each city, a set of geo-targeting queries are submitted to the Flickr API. However, in case the number of geotagged images is not sufficient, additional images will be collected by retrieving images tagged with the city name or tagged with known landmarks in the city.

As far as images captured from the mobile app are concerned although they are handled in the same way as social media images, they are expected to be of much higher quality and consistency since their capturing parameters will be controlled by the hackAIR mobile app.

Finally, webcams can be used as another source of images that depict parts of the skyline of an area of interest. It should be noted that a simple preprocessing of the initial video stream is required to extract frames at specific rates.

4.3 Hardware-Based Sources

These include air quality estimations produced by low-cost open sensor devices assembled by citizens. These estimations will be performed using open hardware and software. More specifically, the widely used Arduino development platform [9] along with the newly introduced PSoC 4 Bluetooth Low Energy (BLE) [28] kits will be programmed for a series of widely available PM sensors in order to enable any potential user to perform and contribute air quality measurements. Preconfigured and open software modules will be provided to users along with suggested hardware configurations in order to enable them to implement low-budget air monitoring stations. Data transmission will take place by means of BLE enabled smartphones and an associated Android application. An additional data collection system oriented to users less familiar with electronics will be used. The proposed system will be built around common off-the-shelf materials. The operation principle is to force air (using aquarium or camping pumps) to pass through a paper filter and after a predefined period of exposure to take a photo (by smartphone) of the filter. Then by using computer vision algorithms, developed specifically to handle such images, the colorization of the filter will be translated to PM estimations. More specific, the colorization of the examined filter that will be caused by PMs will form unique or small clusters on the top layer of the filter. These clusters will be projected in image as blobs. Then a blob detector can be applied that will provide the number of identified and marked blob regions which can be correlated with PM concentration (after calibration). The algorithm has the following steps: Thresholding, grouping, merging and blob radius calculation.

5   Data Analysis

5.1   Web Information Extraction

This module involves the extraction of environmental content from websites and web services. In case of web services the format of the data is well defined and data can be retrieved by simple JSON/XML parsing. As far as websites are concerned, information extraction from environmental websites cannot be realized using deep semantic analysis given that most of the information to be retrieved is not reported in text and often there is little linguistic context available. Thus, information extraction can be based on the extraction and transformation of semi-structured web content, typically in HTML format, into structured data. This task typically involves acquiring the page content, processing it, and extracting the relevant information using XPath and CSS selectors. Selectors are patterns that address elements in a tree. Specifically, XPath [35] is a language for addressing specific elements in an XML or HTML document, while CSS [16] is a style sheet language for adding style in an HTML page. Therefore, XPath and CSS Selectors can select a set of elements or a single element from a tree of objects. The information that is extracted from the websites or retrieved from the service includes the PM measurements, and the area of coverage and date/time the measurements refer to.

5.2   Sky Detection

The module involves two image processing operations: (1) visual concept detection based on low-level feature generation and classification for detecting images that contain substantial regions of sky, and (2) localization of the sky regions within the image. As far as visual concept detection is concerned different detectors will be studied including images representation with SIFT, SURF, aggregation using VLAD and training using Logistic Regression [23]. Another technique involves representation of images using a pre-trained Deep Convolutional Neural Network and training with Linear SVMs [22]. As far as sky localization is concerned, the selective search technique will be tested that combines the strength of both an exhaustive search and segmentation [33].

5.3   Air Quality Estimation

The module involves the estimation of air quality from user-generated photos or webcams. In general, several studies [30, 36] have shown that the color ratio R/G in digital images can be used to derive information about the aerosol content of the atmosphere. Specifically, the system will use the Santa Barbara DISTORT Atmospheric Radiative Transfer Model (SBDART, [29]) to simulate the R/G ratios for a set of solar zenith angles (SZA) and a set of Aerosol Optical Depths (AOD), the latter approximating the PM load. The R/G ratio will be approximated using the ratio of the diffuse irradiance of two wavelengths (550 nm and 700 nm) rather than the radiance. The resulting R/G ratios will be used to create a 3-D lookup table (Table RG) containing R/G, AOD and SZA. Further, the SZA for each day of the year and each hour of the day will be computed for the geographical latitude/longitude of the urban areas of interest and 3-D lookup tables will be created for each urban area of interest containing SZA, Day of Year (DoY) and Time of Day (ToD). Each geotagged image whose coordinates are within the area of interest will be processed for extraction of information on the mean sky R/G ratio that will be computed automatically for the image, the image coordinates, the DoY and ToD the image was taken.

The table for the respective coordinates will be accessed, receiving as input the DoY and ToD and giving as output the SZA. The Table RG will be accessed, receiving as input the SZA and the image R/G ratio and giving as output the AOD, which, together with the image coordinates will be used to plot the AOD value on a city map.

6   Data Storage and Indexing

In order to store and index efficiently the information retrieved from the previously described sources, it is essential to employ a database, which can store efficiently the measurements along with the related information (i.e. date-time, area coverage, and source). Thus, each source can be considered as a sensor, which provides measurements. Hence, one option for handling this information is through a Sensor Observation Service infrastructure, which provides a generic and flexible means for accessing data produced by sensors [24]. This includes access to measurements of the sensors, as well as access to information about the observed features of interest and information about the sensor. The flexibility of the Observation and Reference Model (O&M) can be used for accessing heterogeneous data via a single standard service interface [15]. Another option would be to use a Mongo database that is not oriented towards sensor-based data but on the other hand is more scalable and offers a number of indexes and query mechanisms to handle geospatial information.

7   Conclusions

The hackAIR system builds upon the concept of monitoring and fusing heterogeneous and user-generated air quality monitoring resources towards providing reliable measurements [18]. Towards fusing observational data from the aforementioned sources we plan to evaluate methods based on geostatistics, which build upon previous studies demonstrating its feasibility, such as [17]. In [17], residual kriging is used to combine the sensor observations with a static base map obtained from a geophysical or statistical model. Other fusion techniques that we plan to evaluate include combination of land-use regression techniques with statistical air quality modelling [21]. Eventually the fused data will be used to provide personalised services with respect to environmental issues that will raise the awareness of the citizens on air quality and engage them actively in measuring and publishing air pollution levels. Finally, it should be noted that the system will provide to people near-real time information retrieved from many sources and not predictions like the existing deterministic models do.

Acknowledgments. This work is partially funded by the European Commission under the contract number H2020-688363 hackAIR.

References

1. ActAQ! https://www.newschallenge.org/challenge/data/entries/actaq-an-air-poll ution-mobile-app-with-you-mind
2. Aiello, L.M., Petkos, G., Martin, C., Corney, D., Papadopoulos, S., Skraba, R., Goker, A., Kompatsiaris, I., Jaimes, A.: Sensing trending topics in twitter. IEEE Trans. Multimedia **15**(6), 1268–1282 (2013)
3. Air Visual. https://airvisual.com/
4. AirCasting. http://aircasting.org/
5. AirForU. https://www.uclahealth.org/Pages/AirForU-App.aspx/
6. AirTick. https://www.youtube.com/watch?v=l11abvYgvBY
7. Amsterdam Smart Citizens Lab. https://waag.org/en/project/amsterdam-smart-citizens-lab
8. App for Kids to ID Asthma Attack Triggers. http://www.healthdatamanagement. com/news/App-for-Kids-to-ID-Asthma-Attack-Triggers-51675-1.html
9. Arduino. https://www.arduino.cc/en/Main/Products

10. Banshirne. http://banshirne.com/
11. CAPTOR. http://captor-project.eu/
12. Chen, H., Fan, H., Chau, M., Zeng, D.: Metaspider: meta-searching and categorization on the web. J. Am. Soc. Inform. Sci. Technol. **52**(13), 1134–1147 (2001)
13. CITI-SENSE, http://www.citi-sense.eu/
14. Clean Air Nation (greenpeace India). http://www.greenpeace.org/india/clean-air-nation/
15. Cox, S., et al.: Observations and measurements-xml implementation. OGC document (2011)
16. Selectors Level 3. https://www.w3.org/TR/css3-selectors/
17. Denby, B., Schaap, M., Segers, A., Builtjes, P., Horálek, J.: Comparison of two data assimilation methods for assessing PM_{10} exceedances on the European scale. Atmos. Environ. **42**(30), 7122–7134 (2008). http://www.sciencedirect.com/science/article/pii/S1352231008005803
18. Epitropou, V., Karatzas, K.D., Bassoukos, A., Kukkonen, J., Balk, T.: A new environmental image processing method for chemical weather forecasts in Europe. In: Golinska, P., Fertsch, M., Marx-Gómez, J. (eds.) Information Technologies in Environmental Engineering. Environmental Science and Engineering, vol. 3, pp. 781–791. Springer, Heidelberg (2011)
19. idokep. www.idokep.hu
20. iSCAPE. http://horizon2020projects.com/sc-climate-action/h2020-making-cities-sustainable/
21. Johansson, L., Epitropou, V., Karatzas, K., Karppinen, A., Wanner, L., Vrochidis, S., Bassoukos, A., Kukkonen, J., Kompatsiaris, I.: Fusion of meteorological and air quality data extracted from the web for personalized environmental information services. Environ. Model. Softw. **64**, 143–155 (2015)
22. Krizhevsky, A., Sutskever, I., Hinton, G.E.: Imagenet classification with deep convolutional neural networks. In: Advances in Neural Information Processing Systems, pp. 1097–1105 (2012)
23. Markatopoulou, F., Mezaris, V., Pittaras, N., Patras, I.: Local features and a two-layer stacking architecture for semantic concept detection in video. IEEE Trans. Emerg. Top. Comput. **3**(2), 193–204 (2015)
24. Na, A., Priest, M.: Sensor observation service. Implementation Standard OGC (2007)
25. Oyama, S., Kokubo, T., Ishida, T.: Domain-specific web search with keyword spices. IEEE Trans. Knowl. Data Eng. **16**(1), 17–27 (2004)
26. PASODOBLE. http://lap.physics.auth.gr/pasodoble.asp
27. PESCaDO. http://www.iosb.fraunhofer.de/servlet/is/30549/
28. PSoC. http://www.cypress.com/products/psoc-4-ble-bluetooth-smart
29. Ricchiazzi, P., Yang, S., Gautier, C., Sowle, D.: Sbdart: a research and teaching software tool for plane-parallel radiative transfer in the earth's atmosphere. Bull. Am. Meteorol. Soc. **79**(10), 2101–2114 (1998)
30. Saito, M., Iwabuchi, H.: A new method of measuring aerosol optical properties from digital twilight photographs. Atmos. Meas. Tech. **8**(10), 4295–4311 (2015)
31. Tang, M., Agrawal, P., Pongpaichet, S., Jain, R.: Geospatial interpolation analytics for data streams in eventshop. In: 2015 IEEE International Conference on Multimedia and Expo (ICME), pp. 1–6. IEEE (2015)
32. Ubreathe. http://ee.ricardo.com/ubreathe/
33. Uijlings, J.R., van de Sande, K.E., Gevers, T., Smeulders, A.W.: Selective search for object recognition. Int. J. Comput. Vision **104**(2), 154–171 (2013)

34. World Air quality. https://itunes.apple.com/us/app/world-air-quality/id396958 256?mt=8
35. XML Path Language (XPath). https://www.w3.org/TR/xpath20/
36. Zerefos, C., Tetsis, P., Kazantzidis, A., Amiridis, V., Zerefos, S., Luterbacher, J., Eleftheratos, K., Gerasopoulos, E., Kazadzis, S., Papayannis, A.: Further evidence of important environmental information content in red-to-green ratios as depicted in paintings by great masters. Atmos. Chem. Phys. **14**(6), 2987–3015 (2014)
37. Zheng, H.T., Kang, B.Y., Kim, H.G.: An ontology-based approach to learnable focused crawling. Inf. Sci. **178**(23), 4512–4522 (2008)

ENVI4ALL: Personalised Air Quality Information Based on Open Environmental Data and User-Generated Information

Evangelos Kosmidis[1], Konstantinos Kourtidis[2],
and Panagiota Syropoulou[1(✉)]

[1] DRAXIS Environmental S.A., Mitropoleos 63, 54623 Thessaloniki, Greece
{kosmidis, syropoulou.p}@draxis.gr
[2] Department of Environmental Engineering, Democritus University of Thrace,
Vas. Sofias 12, 67100 Xanthi, Greece
kourtidi@env.duth.gr

Abstract. Air pollution open data has a huge value for citizens, especially these belonging to vulnerable groups. Information on air quality can help them to take better informed decisions that safeguard their health. Although this information is available in multiple sources, in the form that the data is provided, it is difficult for citizens to extract the information they actually need. In addition, existing monitoring stations mainly cover only large cities, and fail to take into account differences in microclimates occurring within a specific area. ENVI4ALL will be an application that addresses these challenges by providing direct access to personalised and localised information on air quality (current, forecast, and historical), making use of diverse sources of large datasets of open air quality data, and crowdsourced information on the perception of app users about the current air quality. An empirical model will be also applied for the provision of air quality forecasts.

Keywords: Air pollution · Open data · User-generated data · Personalised air quality information · Air quality mobile app

1 Introduction

At present air pollution is one of the most significant factors posing threat to health worldwide. According to the World Health Organization, outdoor air pollution was responsible for the deaths of some 3,7 million people under the age of 60 in 2012 [1]. Furthermore, there is ample evidence for various effects of air pollution on different health aspects, including widely common diseases such as asthma, allergies, heart attacks etc. However, according to Flash Eurobarometer 360 [2], nearly six out of ten Europeans do not feel informed about air quality issues in their country (Fig. 1).

Susceptibility to air pollution may differ significantly among individuals as it depends on their personal health profile, habits and behaviour. Personalised information on air quality is crucial for healthy behaviour shaping and promotion, as it enables individuals to plan their activities and take decisions in a way that reduces their exposure to air pollution, and protects their health. Air pollution data is widely

© Springer International Publishing AG 2016
A. Satsiou et al. (Eds.): IFIN and ISEM 2016, LNCS 10078, pp. 157–166, 2016.
DOI: 10.1007/978-3-319-50237-3_8

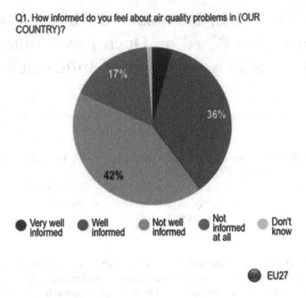

Fig. 1. How informed Europeans feel about the problem of air pollution (source: Flash Eurobarometer 360)

available, as required by the Aarhus Convention (signed by the European Community and its Member States in 1998), which provides for the right of everyone to receive environmental information that is held by public authorities. However, although such information is available, it is not easily accessible by citizens. Air pollution information is scattered in various sources, and the form in which the data is provided is usually difficult to understand. Indicatively, Fig. 2 presents some air quality information for Athens, Greece as it is provided by the Greek Ministry of Environment.

Particularly for historical data, in its current form it is of little value to citizens, as it is not easy to interpret large datasets without using statistical methods. Moreover, environmental information is presented in a generic form, which does not take into

ATTICA STATIONS

	Today on 23/08/2016 the levels until 13:00 varied:	Yesterday on 22/08/2016 the levels varied:
Ozone	• from 1 µg/m³ at the **PATISION**, station • to **144** µg/m³ at the **N.SMYRNI** station	• from **1** µg/m³ at the **PATISION**, station • to **177** µg/m³ at the **MAROUSI** station
	Public information level **180 µg/m³** -- Alarm level **240 µg/m³** . Hourly mean values	
Nitrogen Dioxide	• from 1 µg/m³ at the **THRAKOMAKEDONES**, station • to **111** µg/m³ at the **PIREAUS** station	• from **1** µg/m³ at the **THRAKOMAKEDONES**, station • to **142** µg/m³ at the **PATISION** station
	Alarm Level **400 µg/m³** Hourly mean values.	

Fig. 2. Air pollution measurements in Athens, Greece on 16/8/2016 provided by the Greek Ministry of Environment (source: http://goo.gl/mCL8SE)

account differences among individuals (depending on their personal health profile, habits and behaviour). Another crucial issue is the fact that the measurements from air pollution monitoring stations refer to large scales and fail to define air quality in microclimates occurring within a specific area. Additionally several European countries do not have an efficient monitoring network that measures air quality levels throughout the whole country [2]. Thus, there is an undeniable need of citizens for direct access to targeted, localised and easy-to-understand information on the air pollution levels they are exposed to.

2 Previous Work

There are several initiatives aiming to provide simplified information about health related environmental conditions, with the majority of them citing the problem of air pollution.

A representative national web portal that provides AQI information is the public Air Quality portal of the Department for Environment Food and Rural Affairs of UK[1], which displays air quality index values for UK, as well as historical data of the main air pollutants in a map interface. Other similar national portals are those of Cyprus, Germany and US. On top of these, the AQICN application[2] provides real-time estimations of the Air Quality Index almost for every country around the world. This type of applications focuses on current and forecasted data. Historical data are most often not included, and even if they are, they are provided in the form of data measurements. In this form, it is difficult for citizens to extract the information they actually need, and to identify patterns and trends.

Moreover, two relevant to ENVI4ALL projects are ENVIROFI and PESCaDO. The ENVIROFI[3] project was funded under the Future Internet Public-Private Partnership Programme (FI-PPP) and resulted to a product providing information and measurements on air pollutants, allergens and meteorological conditions. Moreover, users can enter into the system their perception about the current weather conditions and air pollution levels. On the other hand, the PESCaDO project[4] delivered personalized environmental information from multiple sources to users. The information was in the form of indices.

ENVI4ALL, apart from providing information about the current and forecast air pollution levels, it will empower citizens to turn open data into knowledge, helping them to make sense out of complex, large datasets of historical air quality data, extracting relevant information in simple graphs that allow them to identify unseen patterns. However, its big innovation is the fact that it will facilitate access of users to information on health-related environmental conditions, whatever the locations of their interest will be. Even in locations for which there is lack of air pollution measurements,

---

[1] http://uk-air.defra.gov.uk/forecasting/, Accessed: 16/8/2016.

[2] http://aqicn.org/map/europe/, Accessed: 16/8/2016.

[3] http://www.envirofi.eu/, Accessed: 16/8/2016.

[4] http://pescado-project.upf.edu/, Accessed: 16/8/2016.

users will have access to the ENVI4ALL crowdsourcing component and they will be informed about how other people in their vicinity feel about air quality. Thus, they will be able to adapt their activities and behaviour in order to avoid health impairments according to their concerns.

3 The ENVI4ALL Solution

ENVI4ALL will be a mobile application that provides direct access to targeted, localised, and easy-to-understand information on air quality, making use of open data and crowdsourced information based on current environmental conditions, as perceived by its users.

Specifically, ENVI4ALL will offer information on the current, forecast, and historical environmental conditions for a specific location, estimating the value of the Air Quality Index, an index for reporting daily air pollution levels based on the associated health effects people may experience from breathing polluted air.

According to Mintz [3], AQI values for the present day are calculated by an algorithm that takes into account the present day's mean concentrations of a variety of pollutants, such as ozone molecules, nitrous oxides and sulphur dioxide, and determine the AQI value based on the pollutant with the maximum concentration. In this step, we assume that the pollutant with the maximum concentration is the one that affects human health most severely. Based on the daily AQI value, the air quality is categorized in six classes with different colour representation for the best understanding of the public (Fig. 3). Each class is accompanied with health advices for sensitive groups and the general public.

Air Quality Index Levels of Health Concern	Numerical Value	Meaning
Good	0 to 50	Air quality is considered satisfactory, and air pollution poses little or no risk.
Moderate	51 to 100	Air quality is acceptable; however, for some pollutants there may be a moderate health concern for a very small number of people who are unusually sensitive to air pollution.
Unhealthy for Sensitive Groups	101 to 150	Members of sensitive groups may experience health effects. The general public is not likely to be affected.
Unhealthy	151 to 200	Everyone may begin to experience health effects; members of sensitive groups may experience more serious health effects.
Very Unhealthy	201 to 300	Health warnings of emergency conditions. The entire population is more likely to be affected.
Hazardous	301 to 500	Health alert: everyone may experience more serious health effects.

Fig. 3. AQI classes (source: https://airnow.gov/index.cfm?action=aqibasics.aqi, Accessed: 16/8/2016)

ENVI4ALL users will have the option to follow one (or more) out of 6 distinct groups of interest, namely (i) patients suffering from respiratory and/or cardiovascular conditions, (ii) people with outdoor activities, (iii) parents of young children, (iv) elderly/companions of elderly and (v) pregnant women (iv) general public. Based on the profile selected, they will have access to historical, current and forecast air quality information for specific locations of interest, while they will receive personalized recommendations on how to avoid exposure to hazardous air pollution, as well as alerts in case of high air quality levels. In this way users will be able to make evidence-based decisions on their activities that allow them to shape healthier behavioural patterns and thus improve their quality of life.

Another feature of ENVI4ALL is the provision of information on the forthcoming levels of air pollution so that users can act proactively to safeguard their health and wellbeing. Specifically, the forecasting procedure is considered as a nonlinear regression problem between predictors (weather forecast and air quality variables) and predict and (hourly concentration of the pollutant), which is solved by using an intelligent algorithm (neural networks, bayesian networks, support vector machines etc.). The intelligent algorithm is trained by using time series of air quality and meteorological observations of the past two years for any area of interest, using the 12 UTC GFS forecast for the next 24 h as input weather forecast data, to simulate the hourly concentration distribution of the pollutants for the next day. According to the forecast conditions, ENVI4ALL will provide users with alerts and activity recommendations in a language that addresses their specific needs. It is considered that this methodology will be patented, thus little information can be disclosed at this stage.

Although ENVI4ALL is under development, a first version of the mobile application has been launched for two Greek cities, namely Thessaloniki and Athens, for validation purposes so that valuable feedback from users is acquired. This version includes all the planned features except from the provision of air quality forecasts. More cities all around Europe will be included in the next versions.

4 The Technology Behind ENVI4ALL

4.1 Information Flow

ENVI4ALL will be a system that offers personalised and localised alerts/information based on air quality data which is being built upon users' information on their perception of the outdoor air quality and ready measurements provided as open datasets. In a way, ENVI4ALL will transform the stakeholder to both information receivers and information providers in order to complement or improve air quality measurements. The main aim is to shape and promote healthy behavioural patterns as it will provide users with personalised information which they can use in order to take better informed decisions that safeguard their health and on the same time reduce their exposure to air pollution through targeted, localised, and easy to understand information about air quality.

The released version of the ENVI4ALL mobile application is built upon HTTP technologies and uses REST APIs for the communication between 3rd party and own components. An ENVI4ALL information flow can be seen in Fig. 4:

– Data sources: Environmental data for air quality is fetched both from open data and using crowdsourcing techniques. Open data refer to concentrations of air pollutants and are acquired from the official air quality monitoring stations that are already installed in various cities around the world. These data are constantly checked for their variability by the responsible municipality/prefecture and updated automatically in the ENVI4ALL application. User generated data refer to users' perception on the current air quality: users are asked once a day to characterize the air quality in their current area ("Good", "Moderate", "Unhealthy", "Suffocating"). The perceptions of all users are presented on the map interface of the mobile application. User actions/behaviour (e.g. the hour the app was accessed, the time spent in specific parts of the app, the health suggestions that were read etc.) are collected in order to be analysed and extract conclusions that offer a more personalised user experience.

– Analysis: Contains rules that adapt runtime environment to user's needs, and models that transform user's perception of air quality to measurements, either to fill in an incomplete dataset or to validate/update existing measurements. A behaviour tracking model combines user actions with user personal data (location, personal profile, habits and observations) to adjust the application's UI and user's interaction in a personalised way, so that each user gets meaningful information according to their needs/preferences. Lastly an air quality monitor module prepares personalised recommendations and alerts on "poor" air quality incidents based on predefined rules.

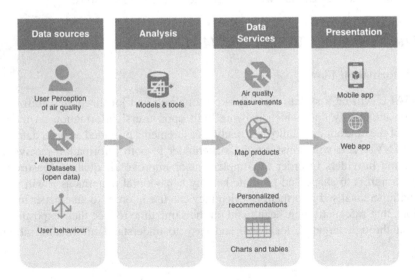

Fig. 4. ENVI4ALL information flow

- Data Services: ENVI4ALL provides RESTFUL services to the mobile and web application and to any other system that might need to consume them. The services oriented architecture (SOA) logic that ENVI4ALL is built upon provides various outcomes like adapted air quality measurements, user recommendations and alerts.
- Presentation: It is the environment that user interacts with, during runtime. This interaction is monitored and logged (behaviour tracking) in order to be fed again in ENVI4ALL and generate meaningful information that helps at personalising the app to fit each user. Several visualisation controls are used to present the content in a visually stimulating way, and enable users to identify trends, and their evolution over time.

4.2   Architecture

ENVI4ALL's architecture is based on the principles of modularity, flexibility and performance optimization. As seen in Fig. 5, it is composed of three basic layers:

Data Layer. Data are stored and retrieved from in-house databases and files, or third party resources such as open data providers, crowdsourcing, authorisation services etc. The data types used at this moment are:

Air quality data from:

- Open Data: An open data crawler service was developed, in order to fetch and store air quality measurements available as open data sources.
- Crowdsourcing: Users are prompted to publish their perception about air quality, in order to enhance the accuracy of information. Users' observations are managed by

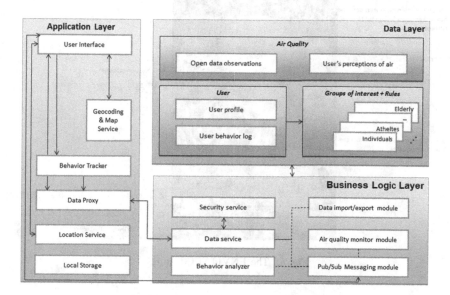

Fig. 5. ENVI4ALL architecture layers

ENVI4ALL RESTful API. However, as these observations only refer to the perception of the users on the current air quality and cannot be quantified, this information is not integrated with the air quality measurements that are acquired from the official monitoring stations. The users' perceptions are only presented on the map interface of the ENVI4ALL application.

User data:

– Profile Data: It is gathered from user's inputs on creating/updating their account profiles. ENVI4ALL RESTful API is responsible to fetch, validate and store the information into the main storage.
– Behaviour observations: System tracks and stores user's interaction with the application interface, in order to use this information for further personalised presentation and related notifications.

Business Logic. The business logic layer responds to requests following the application logic and replies securely with proper content to the client applications defined later in the application layer.

Application Layer. The application layer hosts the user's interaction interface. It is available both as mobile and web application. This layer offers full end-user access to all ENVI4ALL services (Fig. 6).

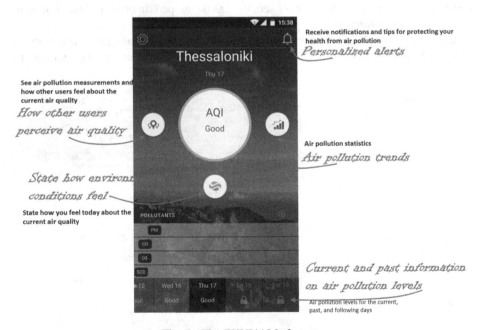

Fig. 6. The ENVI4ALL features

5 Validation by Real Users

Individual users who are concerned about air pollution, and particularly susceptible people, are the final users of the ENVI4ALL application. ENVI4ALL is being built for them so as they are able to protect their health from air pollution, and make evidence-based decisions on their activities that will allow them to improve the quality of their life. In order to increase the marketability of the ENVI4ALL service, the application was validated by real users through open interaction sessions.

The feedback was acquired with questionnaires that were distributed to 50 individual users. This questionnaire can be access through the following URL: http://goo.gl/RQha9L. Although ENVI4ALL is planned to be sold in businesses and public authorities, individual users were asked about their willingness to buy such a solution only to explore their interest on the application. Some findings are presented below.

The majority of the respondents were male (52%), while the dominant age group was 25–34 (37%), followed respectively by the 35–54 (33%) and 18–24 (20%) groups. Less respondents aged above 54 while none of them was younger than 17 years old. Regarding their educational background, the vast majority of the respondents were at least university graduates (85%), while regarding the ENVI4ALL target groups, the majority of them exercises outdoors (41%) or they are parents of young children (39%). Individuals with respiratory and/or heart diseases represented 14%, elderly or companions of elderly represented (4%), while pregnant women (2%) were relatively less.

According to the validation findings the vast majority of the respondents (64%) found it fairly or very easy to navigate the ENVI4ALL application, while 68% of all the respondents found the application visually appealing and 54% the provided personalised information useful. Moreover, some respondents took the initiative to express their overall impression from the use of ENVI4ALL, something that positively surprised the ENVI4ALL team. The majority of the respondents stated that they were satisfied that they were able to choose multiple groups of interests, while some others pointed that the notifications worked properly for them providing specific information for their needs. Finally, a promising business prospect for ENVI4ALL was revealed and stems from the reported willingness of the respondents to buy ENVI4ALL.

6 Conclusions and Further Work

ENVI4ALL will be a mobile application that addresses the need of individuals to receive personalised information on the current, forecast and historical air pollution levels for the locations of their interest. To do that it will harness the benefits of open data on air quality acquired by established public air quality monitoring stations, as well as user-generated data on user's perception about the current air quality.

Except from the marketing opportunities, ENVI4ALL can shape healthy behavioural patterns as it will offer each user with personalised recommendations and instructions for their health protection from air pollution. In addition, the collection of crowdsourced data on perceptions on air quality is expected to create large datasets of real life condition in the cities and will help in the improved understanding of the characteristics of air pollution. This knowledge will be of great importance, in the long term, for research purposed, but also in policy making in the fields of urban planning, public health, transport planning and environmental protection.

Future work will focus on integrating additional data sources for more cities not only in Greece but all around Europe, and on providing air quality forecasting using an empirical model which is developed by the ENVI4ALL team.

References

1. World Health Organisation: 7 million premature deaths annually linked to air pollution. News release, 25 March 2014
2. Andrews, A.: The clean air handbook-a practical guide to eu air quality law. The Clean Air project. Version 1.0, April 2014
3. Mintz, D.: Guidelines for the reporting of daily air quality – the air quality index (AQI). U.S. Environmental Protection Agency, EPA-454/B-06-001 (2006)

Author Index

Printed in the United States
By Bookmasters